HOMER W. SMITH is one of the outstanding figures in modern physiology. Presently Chairman of the Department and Professor of Physiology at the New York University School of Medicine, he has undertaken extensive researches on the structure and function of the kidney, and has written, in addition to numerous scientific articles, the following key works on the subject: *The Physiology of the Kidney; The Kidney: Structure and Function in Health and Disease;* and *Principles of Renal Physiology.*

Born in Denver, Colorado, in 1895, he prepared for a career in medicine at the University of Denver where he received his A.B. degree, and went on to Johns Hopkins University for his Sc.D. degree. As a National Research Council Fellow in Physiology, Dr. Smith studied for two years at the Harvard Medical School. A recipient of the Passano and Lasker awards, and a former Guggenheim fellow, he has also been awarded the Presidential Medal for Merit.

Dr. Smith is a member of the National Academy of Sciences, the American Physiological Society, the American Society for Biological Chemists, the Association for American Physicians, the Society for Experimental Biology and Medicine. Formerly president of the Mount Desert Island Biological Laboratory, he is presently a trustee of the Bermuda Laboratory for Biological Research.

He is also the author of *Kamongo: The Lungfish and the Padre*, a novel; *The End of Illusion*, and *Man and His Gods.*

FROM FISH TO PHILOSOPHER

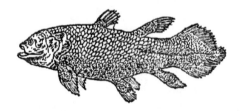

HOMER W. SMITH

WITH 11 ILLUSTRATIONS

PUBLISHED IN CO-OPERATION WITH
THE AMERICAN MUSEUM OF NATURAL HISTORY

THE NATURAL HISTORY LIBRARY
ANCHOR BOOKS
DOUBLEDAY & COMPANY, INC.
GARDEN CITY, NEW YORK

`

The Natural History Library Edition, 1961
by special arrangement with Little, Brown and Company
and CIBA Pharmaceutical Products Inc.

FOR MARGARET AND HOMER WILSON SMITH

ACKNOWLEDGMENTS

Grateful acknowledgment is due the YALE UNIVERSITY PRESS for permission to use the quotations from Dr. George Gaylord Simpson, pp. 15–17, 175; to the *Illustrated London News* for permission to reproduce Figure 11, and to many colleagues, particularly MR. JAMES W. ATZ, DR. FRANK A. BEACH, DR. C. M. BREDER, DR. GEORGE L. CLARKE, MR. CHRISTOPHER COATES, DR. GEORGE GAMOW, DR. CHESTER W. HAMPEL, DR. DONAL SHEEHAN, DR. GEORGE GAYLORD SIMPSON, MR. PETER SMITH, and DR. S. BERNARD WORTIS for advice on technical matters.

The writer is also indebted to MR. PHILIP S. MILLER, Acting Chief of the Music Division of the New York Public Library, for Sir James Paget's account of the Mendelssohn *presto*; to MR. HAROLD C. SCHONBERG, of the Music Department of *The New York Times*, for calling his attention to the Godowsky-Chopin 'Badinage'; and to MR. ABRAM CHASINS, MR. HAROLD LAWRENCE, MR. SIGMUND SPAETH, MISS ROSALYN TURECK, and MRS. NATHANIEL W. LEONARD for numerous suggestions. He is particularly grateful to DR. DAVID SAPERTON for aid in the analysis of music presented here, and for making available for counting his copies of Godowsky's music.

Special acknowledgment must be paid to MISS MARY LORENC for her careful execution of the illustrations.

FOREWORD

The way the kidney works is described clearly and explicitly in this book—a story of vertebrate evolution and adaptation seen through kidney function. The author, Dr. Smith, is an eminent kidney physiologist whose broad range of interests takes him to the realm of evolution—without ever leaving the kidney. He describes how the kidney functions on different phylogenetic levels and how it facilitates the adaptation of organisms, such as the estivating lungfish ("Kamongo" of an earlier book), to highly specialized environments.

The importance of the kidney cannot be overestimated; it removes and excretes the waste products of cell metabolism and it maintains, in constant and relatively equivalent composition, the body fluids which bathe the cells. The kidney works closely with the blood system, removing the wastes from it and adjusting the level and balance of sugars, salts, and water as the blood courses through the fine structure of the kidney. In fact, its basic unit, the nephron, is made up of a tuft of intertwined blood capillaries (the glomerulus) closely aligned with a kidney tubule. Through mechanisms of filtration and reabsorption, the balance of the internal environment is maintained. This constancy of environment is critical to vertebrate evolution and adaptation.

In addition to his comprehensive survey, Dr. Smith deals with the problem of the origin of the earliest vertebrates, ancient fossil fishes called the ostracoderms, which appeared in the fossil record 450,000,000 years ago. Here he postulates that the first fishes appeared in fresh water because the kidney of primitive types of fishes living today is adapted to life in fresh water. It has a large glomerulus, and therefore it is able to filter and excrete the large quantities of water which continuously dilute body tissue. When the book was first published in 1953, paleontologists agreed that the first fishes probably appeared in a fresh-water habitat, that Dr. Smith's evidence was critical, and that it strengthened the paleontological findings: the fauna, flora, and chemical composition of certain geological beds suggested that they were laid down in fresh water. But now the fresh-water habitat theory is no longer completely acceptable. Newly discovered facts about other geological beds, their fauna, flora, and chemical make-up, and the time in which they were laid down, suggest that the first fishes appeared in a marine habitat. Recently at Harvard, where A. S. Romer leads the exponents of the fresh-water habitat theory, Dr. Smith was asked about the evidence presented by the Chicagoan, R. H. Denison, in support of the salt-water theory. Dr. Smith replied that he had read Denison's work carefully, admired it, and felt it merited serious consideration. However, he maintained that his loyalties remain with his Cambridge colleague, Dr. Romer. These two views are in strong opposition, and they will force important revisions when more facts are gathered and analyzed. The problem is intriguing and one that needs solution. Students of paleontology are now working on the problem; this volume should be of considerable importance to their studies.

EVELYN SHAW
Research Associate in
Animal Behavior

March 1961
The American Museum of
Natural History

CONTENTS

ILLUSTRATIONS

Figure 2 in the center of the book shows in synopsis the evolution of the vertebrates in relation to a salt-water (darkly shaded) and fresh-water (lightly shaded) habitat. The irregular curve illustrates mountain-building episodes (geologic revolutions) which have importantly influenced this evolutionary history.

The time scale is such that the Pleistocene era (one million years in length) and Recent Time (about twenty-five thousand years) could not be included, and these are merely suggested by the heavy line at zero time.

The entire period encompassed by documented history is only about six thousand years, or one hundred-thousandth of the interval elapsing since the opening of the Paleozoic era (Cambrian period), when fossilized animals first begin to appear in the sedimentary rocks.

AUTHOR'S NOTE

This book is based in part on studies in the comparative physiology of the kidney carried out by the writer during the past thirty years at the New York Aquarium, the Mount Desert Island Biological Laboratory at Salisbury Cove, Maine, the Bermuda Biological Station for Research, the University of Virginia School of Medicine, and, since 1928, at New York University College of Medicine, in later years supported by the Commonwealth Fund and the National Heart Institute of the National Institutes of Health.

Field studies on the lungfish in Central Africa in 1928 and on fresh-water elasmobranchs in Siam and Malaya in 1930 were conducted while he was a Fellow of the John Simon Guggenheim Memorial Foundation. The evolutionary history of the kidney as developed here was first presented in a short essay on that subject delivered in 1939 as a J. L. Porter Lecture at the University of Kansas School of Medicine.

FROM FISH TO PHILOSOPHER

CHAPTER I

EARTH

Nearly a century has elapsed since Claude Bernard, among the most notable of nineteenth-century physiologists, first pointed out that the true medium in which we live is neither air nor water, but the plasma or liquid part of the blood that bathes all the tissue elements. This 'internal environment,' as he later called it, is so isolated from the world that atmospheric disturbances cannot alter it or penetrate beyond it: 'It is as though the organism had enclosed itself,' he said, 'in a kind of hothouse where the perpetual changes in external conditions cannot reach it.' It was Bernard's view that we achieve a free and independent life, physically and mentally, because of the constancy of the composition of our internal environment.

In Bernard's time the chemistry of living organisms was poorly known and afforded only a meager insight into the complexity of the internal environment. But as the modern sciences of biochemistry and physiology have added chapter after chapter on this subject, this new knowledge has only emphasized the importance of his generalization.

Apart from the red and white blood cells and other 'formed' elements, the extracellular fluid (*i.e.*, blood plasma and interstitial fluid) of all animals contains many different organic and inorganic substances. In the organic category are the plasma proteins and many smaller molecules such as glucose, the amino acids (the building blocks of proteins), the lipides (the building blocks of fat), and a variety of vitamins and hormones, all of them necessary to the nutrition and operation of the body. This organic category also includes many waste products on their way to excretion. In the inorganic category are water itself (the major constituent of the plasma and the interstitial fluid, as well as the tissues) and numerous salts as represented by sodium, potassium, magnesium, calcium, chloride, bicarbonate and phosphate in various combinations, as well as oxygen on its way from the lungs to the tissues for utilization, and carbon dioxide on its way from the tissues to the lungs for excretion.

The lungs serve to maintain the composition of the extracellular fluid with respect to oxygen and carbon dioxide, and with this their duty ends. The responsibility for maintaining the composition of this fluid in respect to other constituents devolves on the kidneys. It is no exaggeration to say that the composition of the body fluids is determined not by what the mouth takes in but by what the kidneys keep: they are the master chemists of our internal environment, which, so to speak, they manufacture in reverse by working it over completely some fifteen times a day. When, among other duties, they excrete the ashes of our body fires, or remove from the blood the infinite variety of foreign substances that are constantly being absorbed from our indiscriminate gastrointestinal tracts, these excretory operations are incidental to the major task of keeping our internal environment in an ideal, balanced state. Our bones, muscles, glands, even our brains, are called upon to do only one kind of physiological work, but our kidneys are called upon to perform an innumerable variety of operations. Bones can break, muscles can atrophy, glands can loaf, even the brain can go to sleep, and endanger our survival; but should the kidneys fail in their task neither bone, muscle, gland nor brain could carry on.

Recognizing that we have the kind of internal environment we have because we have the kind of kidneys that we have, we must acknowledge that our kidneys constitute the major foundation of our physiological freedom. Only because they work the way they do has it become possible for us to have bones, muscles, glands, and brains. Superficially, it might be said that the function of the kidneys is to make urine; but in a more considered view one can say that the kidneys make the stuff of philosophy itself.

This is the story of how our kidneys work, and of how they came to work the way they do—which is the story of the evolution of the vertebrates, of which man is the most notable and intelligent species and the only philosopher. But the story of the evolution of the vertebrates is one with the story of the earth, and it is with the story of the earth that we begin this book.

The planet Earth seems fairly solid beneath our feet, but under the examination of the geologist it is revealed to be an inhomogeneous and unstable sphere that has had a most agitated history. Students of cosmogony are not in unanimous agreement how the solar system was formed, but it is generally agreed that the sun and planets had a common origin from interplanetary dust at a date not much over five thousand million years ago, and about a billion years later the earth acquired its present stratified structure. In cross section, this structure presents several more or less distinct, concentric spheres that differ from each other substantially in composition and temperature. (Figure 1.) The outermost sphere, called the crust, varies in thickness from 2 to 3 miles under the deep oceans, to some 30 to 50 miles in the continental masses. Underneath the crust is what is called the mantle, the outer layer of which is perhaps a basaltic rock, while the inner layer is presumed to consist of metallic sulfides and oxides, the entire mantle extending to a total depth of some 1800 miles. The continents, which are composed largely of granite, are lighter than the underlying mantle and consequently they literally float on the surface of the mantle as icebergs float in the ocean, with the greater part of their mass, particularly under the mountain ranges, submerged to depths of 15 to 40 miles. At the center of the earth is a very dense core, thought to consist of iron and nickel. Even these general statements, which are based chiefly on the velocity of transmission of earthquake waves, are matters of debate: some geophysicists refer to an inner and outer core which differ in density; the composition of the mantle, and particularly the depth of the basaltic crust, is a matter of divided opinion; and the composition of the outermost crustal layer underlying the sea basins is not certainly known. The transition from the crust to the outer layer of the mantle is fairly abrupt and is known as the Mohorovičić discontinuity, and familiarly referred to by geophysicists as the Moho. Geophysicists hope to explore the composition of the outer mantle in the future by boring through the crust where it is thinnest under the ocean.

It is more certain, however, that the pressure inside the earth increases with depth: at the bottom of the mantle the pressure is estimated to exceed 1 million atmospheres, at the center of the earth about 4 million atmospheres. The density increases with increasing pressure: the greatest density of rocks at the surface is about 3, at the bottom of the mantle perhaps 5.6, and the metallic core may have a density of 10 to 12; the average density of the earth is calculated by astronomers to be only 5.5, a quite reliable figure which enables the geophysicist to bring his seismologic and other data into a balanced account. As one bores vertically into the earth the temperature increases about 16° F. for every thousand feet, so that in deep wells the boiling point of water is reached at about 7000 feet; it is in this region that the steam is formed which actuates hot geysers. Assuming a uniform rate of increase, the temperatures of mol-

ten rock, 1200° to 1800° F., will be reached at a depth
of some 30 miles, and it is from this depth (or lower)
that there comes volcanic lava, escaping through faults
in the outer crust. Contrary to what one might expect,
however, the temperature of the surface of the earth is
not determined by the escape of heat from the interior
—because the earth's substance conducts heat very slowly
—but rather by solar radiation; while the temperature
of the interior of the earth (below a few hundred feet)
is determined by the decomposition of radioactive elements, a process liberating much more heat than that
gained from the sun. Such has apparently been the case
throughout most of geologic time. In consequence of the
great pressure at the center of the earth the metallic
core has a temperature perhaps not far below that existing at the surface of the sun.

Paleontologists and geologists have sought by a variety
of methods to learn the exact age of the earth, or at least
of its superficial rocks. The method that is now believed
to be most accurate depends on the ratio of certain
radioactive elements, notably uranium and thorium, to
their decomposition products (chiefly various isotopes of
lead): assuming that only negligible quantities of these
decomposition products were present when the rock first
acquired a solid state, the ratio of the parent radioactive
element to its decomposition product indicates the age
of the rock as a solid matrix. From such analyses Arthur
Holmes in 1947 estimated that the oldest rocks studied
up to that date had an age of 3350 million years, the
methodologic error in this estimate not exceeding ± 10
per cent. The figure represents, of course, the time since
the crust solidified to such an extent that mixing of isotopes had ceased. Other data indicate that the total age
of the earth as a planet is about 4500 million years. This
is almost the age of the solar system, as calculated by
several methods, and indeed the galaxy to which our
sun belongs is estimated by some astrophysicists to have
an age of about 5000 million years.

Despite the long period during which the earth has had a hard crust, it has never acquired a truly rigid structure. This is partly because of the high temperatures of the interior, and partly because of the great pressures to which the interior is subjected—for example, at a depth of only 400 miles the pressure is 8,000,000 pounds per square inch. All solid bodies when placed under pressure behave like viscous fluids, and the earth as a whole has this semifluid character. This 'fluidity' is such that the familiar oceanic tides raised by the moon's gravitational force are paralleled by similar tides in the earth's crust, each circuit of the moon raising and lowering continents and seas, deserts and mountain ranges, by some 16 inches.

According to an interpretation first proposed by Sir George Darwin (son of Charles Darwin), the moon was literally torn out of the earth by a solar tide of this nature; this fission hypothesis goes far to explain several otherwise peculiar facts, but contemporary cosmogonists favor the belief that the earth and moon were formed simultaneously. In any case, as George Gamow has pointed out, if one assumes a uniformly thick crust of granite the volume of water in the oceans is sufficient to cover the entire earth to a depth of several miles. Were it not for irregular distribution of granite in the continents, the evolution of terrestrial life could never have occurred. Throughout most or all of geologic time the surface temperature of the globe has been below the boiling point of water and its surface therefore largely covered by water.

Scarcely less plastic than the interior of the earth are the granite continents themselves, which throughout earth history have been repeatedly warped and wrinkled and at points actually overfolded in mountain chains. The causes of orogeny (oro = mountain; gignesthai = to be born) are multiple and extremely complex. Most important, perhaps, is the heat generated by radioactivity in the interior: this accumulates for a time and leads to

melting and expansion of the subcrust, perhaps in local-
ized areas of convection, with resulting elevation of the
overlying rocks, then the heat is lost from the interior
by volcanic action and other routes, this process giving
rise to periodic 'blistering' of the surface. Under the tidal
pull of the sun the continents have tended to drift westward, and it still is believed by some that at an early
time there was one great land mass that split into fragments which, drifting in periods and areas of high mantle
temperatures, produced the continents as we now see
them. These continents even now are under great strain
which is relieved by fracturing of the thinner parts of
the crust, especially under the oceans, permitting extru-
sion of material from the interior and movements of the
exposed land masses into new positions of equilibrium.
Possibly the earth has contracted during its planetary
history by as much as 200 to 400 miles, wrinkling the
crust like the skin of a drying apple.

Whenever land masses were elevated, frost, ice, sun,
wind, and rain began their work of wearing them down,
weathering them first to boulders and pebbles, and finally to silt that the rivers carried into the sea and deposited on the ocean bottom along the continental ledges.
This silt, in places accumulating to a depth of many
miles, added its weight to that of the overlying water
and pressed down on the underlying basalt and forced
it under the lightened continental masses, floating them
higher above the sea. (A fair idea of the magnitude and
speed of erosion is afforded by the facts that the Colorado River carries annually into the sea an average of
184 million tons of silt; the Mississippi, an average of
730 million tons.)

In any case, the sequence of mountain building followed by erosion has gone on throughout earth's history,
not in a uniform manner but with marked intermittency.
Erosion is rapid when mountains are highest, and slow
or negligible during periods of low continental relief, and

consequently changes in continental weight follow an irregular cycle. Climatic changes, which are themselves
related to the elevation of great mountain ranges, have
at times led to the formation of great glacial ice fields
that have at once lightened the oceans and added their
weight to that of the continents, causing the covered regions to sink deeper into the basaltic layer. Between these
and other factors, mountain building has reached peak
activity at intervals of roughly thirty million years (as
illustrated in Figure 2), with three major episodes separating the Proterozoic, Paleozoic, Mesozoic, and Cenozoic eras, all of which have had dramatic consequences
in the biologic history of the earth.

Major periods of mountain building, the historical
geologist calls 'revolutions'; the lesser periods he calls
'disturbances' (the Rocky Mountains, Alps, and the Himalaya are products of the relatively recent Laramide
and Cascadian 'disturbances'), and the intervening periods of quiescence he calls 'intervals.' During the intervals the mountains raised in preceding orogenic episodes
have been largely if not entirely worn away by wind and
rain: it is estimated that, in the extreme, the total continental depth eroded and carried into the seas since the
opening of the Paleozoic era—that is, in only one-sixth
of the total history of the earth—would, if superimposed
in an unbroken vertical column, exceed ninety-five miles,
or thirty times the present Alps or Rockies. Throughout most of earth's history, however, the general distribution of continents and oceans has been much the
same as it is today.

Had the earth cooled with no mountain-building episodes, it would today be almost entirely covered by
oceanic waters from which there would protrude only
a few widely separated granite islands of relatively low
relief. Life would probably have been evolved in this
geographic wilderness of small islands in a universal sea,
but it is improbable that it would have taken the form

of the thousands of invertebrates that now exist in the sea, because their evolution has been conditioned by the continuous accumulation in sea water of the large quantities of salts leached out of ninety-five vertical miles of the earth's crust. It is even less probable that it would ever have taken the direction of vertebrate evolution and produced the air-breathing fishes, the terrestrial four-footed animals, or the warm-blooded birds and mammals. Except for mountains (and perhaps the moon) life, if any, would be very different—how different can be left to the speculations of the more imaginative writers of science fiction.

We can confidently say that earth history would have pursued its course without the aid of any living organisms. But the history of living organisms has been shaped at every turn by earth's vicissitudes, because every geologic upheaval, by causing profound changes in the distribution of land and sea, has had profound effects on the climates of both, and hence on the patterns of life in both.

Climate has always been much as it is now—better and worse. It is determined by many complexly interacting factors, among which winds and oceanic currents are probably the most important. Astronomical factors, such as periodic changes in the earth's orbit and angle of rotation, have played a part, but these are short-term factors and at their peak probably have contributed less to climatic changes than have the great cycles of continental elevation. By cutting off moisture-laden winds, by diverting oceanic currents, and in many other ways, these mountain-building episodes have repeatedly modified the climate of almost the entire surface of the globe. Periods of maximal continental elevation have generally coincided with periods of maximal glaciation, in which the thick ice extended in places almost down to the sea in what are now temperate or even tropical latitudes—periods which were accompanied by a decrease in the temperature of the oceanic water and withdrawal of marine life toward the equator and the extinction of many forms of plant and animal life. Such periods have also been frequently accompanied by the spread of deserts over large continental areas. On the other hand, periods of low relief between mountain-building episodes have seen the spread of warm, shallow seas across the continents and the extension of tropical oceanic currents into the polar regions, with such amelioration of climate that marine and terrestrial forms of life, now limited to the equatorial regions, have grown abundantly close to the Arctic zone.

The history of the earth and the history of life are related as a die and a mold: as geologic revolution has succeeded revolution, each has left its imprint in the pattern of life as well as in the pattern of the rocks. Living organisms of ever-increasing complexity, and with ever-increasing capacity to lead a free and independent life, are a product of earth's troubled history.

CHAPTER II

EVOLUTION

When Charles Darwin published his epoch-making work, *On the Origin of Species by Means of Natural Selection*, one hundred years ago, he had grasped the two essential processes underlying organic evolution: biological variation and natural selection. Our knowledge of both processes, however, has been greatly enhanced since Darwin's time. It was not until the early decades of the present century that biologists clearly distinguished between acquired characters, or those that are acquired during the life of the individual, and genetic characters, *i.e.*, those that are inheritable generation after generation. There is no evidence that acquired characters are ever inherited, and there is much positive evidence to refute the view. Consequently the raw materials of evolution must be sought in changes in the genetic characters.

Early in this century it was shown that genetic characters are determined by the chromosomes of the cell, and it has since been established that each chromosome consists of many units called genes (from *gignesthai* = to be born) arranged in a definite sequence along the chromosome. Each gene has a very profound influence on one or more body characters, and a 'genic map' of the chromosomes can be made by observing the behavior of these characters in heredity. In general, genes are quite stable and are passed on from one generation to

another without change during the union of the reproductive cells from the parents.

Through the mechanism of sexual reproduction
(which is nearly universal in both animals and plants)
the offspring receives genetic factors from both parents,
and thus differs genetically from either parent, though it
does not necessarily possess any new genes. In this type
of reproduction, variation is limited to shuffling and reshuffling of the parental genes. By repeated brother-and-
sister mating it is possible to obtain males and females
the genetic composition of which is so similar that the
offspring is highly constant in character, giving so-called
'pure-bred' strains—in other words, strains showing minimal variation in body characters. This type of breeding,
however, rarely occurs in nature, because the father and
mother are generally of different genetic strains, and under these circumstances the potentialities for variation
afforded even by the simple reshuffling of genes are almost unbelievably large. For example, with only 23
chromosomes in the human paternal and maternal germ
cells, the total number of different combinations of
chromosomes in the potential offspring of one man and
one woman is over 70 thousand billions, or 26,000 times
the population of the earth. But every chromosome is
composed of many genes, each of which may have several variations: on the assumption that there are a thousand distinctive genes in the human chromosomes (and
there are probably many more), and that each gene has
only two variations (and there are probably more), it has
been calculated that the number of potential combinations of genes to be derived from one man and one
woman is 2 to the 1000th power, a figure greater than
the number of electrons in the world. With random
mating such as occurs in nature, it can therefore be
safely said that no two individuals in any species are ever
exactly alike. This even applies to identical twins, which
may, but do not necessarily, start out with an absolutely
identical genetic pattern but are now thought to undergo

some genetic differentiation during embryonic development.

Though no knowledge of either genes or chromosomes
was available in Darwin's day, it was known to him, and
to plant and animal breeders generally, that occasionally
a carefully inbred stock produced a 'sport' differing markedly from the parents in one or another specific character, and that the new character was stable in that it was
inheritable. This sudden and random appearance of new
varieties was first called 'mutation' by the Dutch botanist,
Hugo De Vries, in 1901, who had studied it carefully in
the American evening primrose (*Oenothera lamarcki-
ana*). De Vries believed that it was such sudden changes
that gave rise to new species and afforded the raw materials of evolution, but it was subsequently shown that
his mutations among primroses were the result simply
of 'chromosome mutations,' which may take the form of
a change in the number of chromosomes (a phenomenon
that occurs more frequently in plants than in animals),
of a rearrangement of genes within a chromosome, or a
transfer of a gene from one chromosome to another. If
De Vries was wrong about the basic nature of his primrose mutations, he was right in principle in surmising
that sudden—or as he called them, 'explosive'—changes
can occur in the basic determiners of hereditary characters, the genes; and his term 'mutation' is applied today
not only to chromosomal mutations but also to sudden
transformations in the genes themselves. Such genic mutations produce new body characters which, so long as
the new gene remains stable, are inherited according to
the laws of genetics. Genic mutations have been inten-
sively studied in many species of animals and plants and,
along with chromosomal mutations, are conceived to
supply the 'variations' which afford the raw materials of
evolution.

Mutations are said to be 'spontaneous,' in that they
occur independently of all known environmental influences. It is known that high energy radiation of cells in-

creases the rate of mutation, but whether the mutations
so produced are those that might arise spontaneously is
not determined. It has been inferred that cosmic radiation may in part be responsible for 'spontaneous' mutation, but the idea is as yet without support in evidence.
A few chemicals such as colchicine and the nitrogen
mustards also produce mutations, probably by reacting
chemically with the chromosomes, but again there is no
evidence that such compounds are responsible for the
naturally occurring process. Consequently when we call
mutations 'spontaneous' we are only admitting that the
underlying causes are unknown.

Mutations are also said to be 'random' in nature:
though limited in quality and number, they may take
any one of several forms, and in the great majority of
instances they have no adaptive value to the organism
and may even be lethal. If a mutation is useful, it is so
by chance alone. Hence, while random mutation supplies
the raw materials of evolution, mutation alone would
be of little avail in modifying organic pattern in the direction of better adapted or more complex animals
were it not for the supplementary operation of natural
selection.

For Darwin, natural selection was a rather crude process of competition between the members of a species,
which worked to eliminate the 'unfit' and to permit the
survival of the 'fit' in the perpetual struggle for existence.
But modern biology sees in natural selection a far more
complex process, and one that begins to operate between
the genes and chromosomes from the instant when the
sexual cells unite in fertilization. It has perhaps completed a large part of its work in the embryonic period,
determining the issue of death or survival of the developing embryo early in the period of gestation. And in so
far as it carries on into adult life, natural selection is
much more complex than Darwin supposed: it goes beyond competition in the grosser sense and includes competence to meet environmental extremes of heat and

cold, the utilization of available water or food with maximal efficiency, a larger number of young or, alternatively, better care of the young and a longer breeding life, optimal co-operation within the group or between different groups that can aid each other, and the exploitation of new environmental possibilities that are not areas of competition.

Natural selection not only fosters the survival of new forms with superior endowments but it drives new forms into special niches of environment in which their unique characters are most advantageous, and by this isolation it prevents them from back-breeding with unmutated forms, a process that could lead to the extinction of their unique genetic characters. And when mutant forms are isolated in special habitats, natural selection determines what new combinations of genes and chromosomes will be available for further mutation.

Where once it seemed that random mutation and natural selection could scarcely in an infinity of time fashion the exquisite patterns of organic adaptation everywhere evident in nature, it is now accepted that, given known or probable mutation rates as determined from living animals and plants, or as judged by the rates of appearance of new forms in the past, the evolution of highly complex adaptations is not only a matter of high statistical probability but one of relative rapidity. Modern students of evolution see natural selection not as blind but as creative and constructive.

George Gaylord Simpson gives an analogy which, though oversimplified, illustrates both the quantitative aspects of modern evolutionary theory and the creative force of natural selection itself:

Suppose that you have in a barrel many copies of all the letters in the alphabet, and that you try, by drawing three letters at a time, to spell the word 'cat,' discarding all three letters every time you fail. You may never succeed, because you may draw all the c's, a's or t's in wrong combinations and discard them without ever hitting the desired combination. But suppose that every time you draw a *c*, an *a*, or a *t* in a wrong combination, you put these desirable letters back into the barrel and discard only the undesirable letters. Your success is now assured, because in time there will be only *c's*, *a's* and *t's* in the barrel; and, indeed, because of their rapid accumulation, you will probably succeed in drawing *cat* in one set of three letters long before the barrel becomes a pure pool of these three letters. Suppose, moreover, that in addition to returning the *c's*, *a's* and *t's* to the barrel (discarding all other letters), you clip together and return any two desirable letters when they happen to be drawn at the same time; you will then accumulate the combinations *ca*, *ct*, and *at* at a rapid rate while you still have a large number of *t's*, *a's* and *c's* needed to complete the combination on a single drawing. Obtaining *cat* now becomes a matter of such high probability that you will succeed long before all the letters in the barrel have been drawn once. Moreover, from the letters given at the start you will have created the word *cat* which did not exist in the barrel at the beginning. Simpson's alphabet is illustrated in principle, but by a different graphic analogy, in Figure 3.

Add to the reshuffling of 'stable' genes in sexual reproduction the facts that, so far as is known, every gene can itself undergo mutation to produce a new gene with markedly different effects on the body (and frequently with profound effects on the action of other genes), and that useful mutations and useful combinations tend to be preserved and 'clipped together' by natural selection, and thus to be returned to the 'barrel' where they supply new combinations for variation and selection to work upon, and it can be seen why evolution is a 'creative' process, working frequently, if not constantly, toward more and more elaborate adaptation. Add the succession of day and night, of summer and winter, add the great cycles in the climate of the sea and earth that have marked the geologic past, and one sees the process of evolution in proper perspective. As Simpson says, 'This process is natural, and it is wholly mechanistic in its operation. This natural process achieves the aspect of purpose, without the intervention of a purposer, and it has produced a vast plan without the concurrent action of a planner.'

The history of evolutionary theory abounds in attempts to show that evolution contains foresight, that it moves toward one or another goal in such a manner that the goal seems to be predetermined. Evolution in a straight line is admittedly discoverable in many instances, but the majority of paleontologists agree that it does not reveal any inherently unidirectional process, much less any metaphysical perfecting principle; the appearance of evolution in a straight line, where it exists, occurs because of the limited variability in mutation and selection and because it is easier to continue a line of evolutionary change, for which many of the determinants are already present, than it is to start off on an entirely new line. Moreover, if the environment is continuously changing in one direction, natural selection tends to give a sustained direction to adaptation. Hence some appearance of design or purpose is inevitable in the over-all process, but we cannot assign to either mutation or selection any foresight, design or plan. To quote Simpson again, 'There is neither need nor excuse for postulation of nonmaterial intervention in the origin of life, the rise of man, or any other part of the long history of the material cosmos.'

We can speak of 'progress' in evolution, but progress is not the essence of it. Aside from the continued filling of the world with highly diverse forms of life (a process which also has had its ups and downs) there is no sense in which it can be said that evolution is intrinsically progressive. It is only 'opportunistic,' in that every conceivable device gets tried and the problem is frequently solved in more than one way. It has been pointed out that the variety of photosensitive cells that have been

developed in various types of animals is almost incredible: it seems nearly impossible to think of a practical photoreceptor that has not appeared in one group
of animals or another, ranging from the diffuse sensitivity
of unspecialized cells to the elaboration of many different types of compound eyes, and to the simple eye of
man—which is not necessarily the best. Again, apart from
the insects, wings have been developed three times
from the forelimbs of the quadrupeds—in the reptilian
pterodactyls, in the birds, and in the mammalian bats—
in each instance representing a different structural experiment and a different solution.

Progress can take a variety of forms, but it is only in
the development of increased physiological independence of environment, which, for mobile forms, involves
increased awareness and perception of the environment
and increased ability to react accordingly, that we can
speak of evolution as being upwards rather than just
sideways All evolution is adaptive, if by evolution one
refers only to those forms that survive beyond the time
and circumstance of their origin; and all evolutionary
adaptation is 'preadaptation' (though students of evolution use this word in a more specialized sense), since
the mutation must come first and the mutant organism
must then get along as best it can. Natural variation is
not adaptive—selection cuts off many more mutants than
by its grace are permitted to survive. It is natural selection that turns the randomness of natural variation into
an organically useful plan.

When Linnaeus (1707–1778), in his *Systema Naturae*
and other works, catalogued the plant and animal king-
doms into phyla, classes, orders, families, genera, and
species, he grouped them (even as Adam did when he
first named the animals) according to those obvious
anatomical affinities with which (so it seemed to him in
pre-evolutionary days) they had been endowed by the
act of Original Creation. But Darwin and all subsequent
evolutionists have seen in these biological relationships

greater or lesser degrees of divergence of animals and
plants from one or more common roots, and have sought
to trace their evolutionary affinity.

At the lowest level of variation—that is, within a species—a population of animals is distinguished by such
minor differences that they are given no classificatory
significance by the biologist, and may present no advantage or disadvantage biologically—as, for example, variation in the numbers of scales and their arrangement, in
coloration or in size, or in breeding or feeding habits.
What does it matter if a fish has 12, 14, or 18 rays in
the dorsal fin except to afford the species splitter the
delight of naming three species instead of one? But if
geographic isolation prevents back-breeding, and mutation continues to increase the number of spines in one
stock and to decrease it in another, the process of variation may lead to two forms that rarely or never inter-
breed and the biologist will be justified in dividing them
into two distinct species. The evolution of such a 'subspecies' within a species is often the beginning of the
evolution of new species.

With more extreme variation the population as a
whole changes its character to such an extent that species become wholly different, giving rise to 'genera,' 'sub-
families,' and 'families.' Here adaptation becomes more
obvious, the randomness of variation having been submerged or obliterated by selection so that evolution begins to take on the appearance of sustained direction.
With further variation and selection, new forms appear
that are now so unlike their predecessors as to give rise
to classificatory units of still higher rank, such as 'orders'
and 'classes.' There is no evidence, however, that even
these major steps involve either a sudden, stepwise transformation or an acceleration of the mutation rate as it is
observed within a species. Rather they are the product
of the long accumulation of small, random mutations
operating in such a direction as to lead to what may be
a distinct change in the mode of life. Here is what the

paleontologist technically calls 'preadaptation': the development of characters that are apparently of no advantage under the circumstances in which they first appear, but that prove to be highly advantageous under new circumstances. In essence, it is simply putting old (and sometimes not very useful) things to new uses under new circumstances.

Hence we cannot speak of those changes of climate which have accompanied the geological revolutions of the past as 'causing' the evolution of new and better-adapted animals—the pressures of natural selection are in no way recognized to be the causes of evolution. And none of the vertebrate classes can certainly be shown to have originated during a period of mountain building: with the possible exceptions of the reptiles and mammals, new classes have appeared in the relatively quiescent intervals separating mountain-building episodes. Nevertheless the pressure of natural selection shapes evolution by turning its course here and there: in the long view, evolution *is* adaptation, and if environmental change did not occur coincidentally or subsequently, to foster or to give advantage to mutant forms, variation would spend itself in vain and new and more highly varied ways of life would have less probability of coming into existence.

Hence it is a valid epitome to say that earth's turbulent history has made us what we are.

CHAPTER III

THE PROTOVERTEBRATE

The first episode of mountain building to leave a recognizable geologic record was the Laurentian revolution, some thousand million years ago, when molten granite of volcanic origin poured over older sedimentary strata, and these strata were themselves lifted and folded into mountain chains. These Laurentian mountains are long since gone but the sediments to which they were reduced remain exposed in areas in eastern Canada and give name to this mountain-building episode. After the Laurentian came the Algoman revolution, then two more, which are obscure, and then the Charnian, which closed the ancient history of the earth—a period covering six-sevenths of its entire duration—and ushered in the Paleozoic (*palaios* = ancient; *zoe* = life) era, five hundred and fifty million years ago, in which life first began to leave indubitable traces of its existence.

Speculations on the origin of life can be deferred until more knowledge is available about the chemistry of the ancient seas. There need have been nothing unique about the circumstances attending its birth except that protoplasm spun itself out of sunlight and the available stuffs of water and air, and operated according to the laws of thermodynamics—with only this to distinguish it from sea, wind, and rain: that by a favorable concatenation of molecular forces it automatically became organ-ized in such a manner as to preserve and to reproduce itself, and even this process presents considerably less difficulty theoretically than it did ten years ago.

The first forms of life—and they may have been highly diversified—were certainly below the cellular level, because even the unicellular animals or plants, or Protista, represent an advanced stage in evolution in which the organism has acquired elaborate mechanisms for nutrition and reproduction. Since most of these unicellular organisms are soft-bodied, it is not to be expected that anything more than traces of their existence will be found in the fossil record. Indeed, few fossil vestiges of life of any kind are discoverable before the Paleozoic era; the most abundant are deposits of calcium-containing algae, with only rare filmlike fossils which might be the remains of bits of seaweed. That other plants, probably of microscopic size, were also fairly abundant is indicated by the presence of carbonaceous matter, in the form of coal or particles of carbon in sedimentary deposits, but what forms they may have had remains a mystery. A few trails or burrows, presumably made by wormlike creatures living in the mud, present the only evidence of animals.

Then, with the opening of the Cambrian, the first period of the Paleozoic era—and the first period shown in the geologic sequence as illustrated in Figure 2—living organisms suddenly appear throughout the seas in great variety and already diversified into all the major branches of the marine invertebrates: the Foraminifera, sponges, coelenterates, worms, brachiopods, gastropods, echinoderms, and arthropods. Whether these invertebrates were evolved from a single primordial form or from several forms that from the beginning had developed independently cannot be said. Given the composition of the sea and air, the nature of available atoms and molecules, they would in any case have pursued a closely parallel course.

The sudden enrichment of the fossil record in the Cambrian is probably attributable not to any abrupt acceleration of evolution but to the development of easily fossilized, hard shells composed of chitin or calcium carbonate. The evolution of hard shells cannot be attributed to any sudden increase in either the calcium or the carbonate content of the Cambrian seas, which probably had much the same composition as the sea today, because both calcium and carbonate had long been abundant, and many of the Cambrian forms alternatively made their shells of chitin, a nitrogenous material resembling the stuff of fingernails and not requiring lime, and still used by the modern insects and crustaceans (the lobster, crab, etc.) to encase their bodies. It is more likely that the accelerated evolution of the hard, external skeleton reflects the spread of the carnivorous habit: whereas free-floating, soft-bodied and defenseless animals that fed on microscopic plants could survive without armor in the sparsely populated sea, they began to eat each other as their numbers and varieties increased, and the advantage accrued to those that had developed armor—which served to protect them against their fellows. It is not surprising that simultaneously with the appearance of protective armor, the invertebrates should have evolved destructive mouth parts, prehensile limbs, and, above all, increased strength and mobility to pursue their prey. The most advanced creatures in respect to swimming, crawling and predaceous or scavenging habits were the segmented eurypterids (see bottom of Figure 4), scorpionlike creatures allied to the present-day horseshoe crab, *Limulus*. These animals made up fully 60 per cent of the known Cambrian fauna, and, by the multiplication of their jointed legs, demonstrate how the ability to move about became, if not a *sine qua non* for survival, certainly a distinct advantage.

About one million species of living animals have been described, of which those with backbones—the vertebrates—comprise only 25,000 species distributed among the fishes, Amphibia, reptiles, birds, and mammals—with man, of course, as a single mammalian species mentioned modestly at the end. We are concerned here only with the vertebrates, because man is a vertebrate. That the lowest vertebrates were themselves evolved out of some invertebrate race is accepted by everyone; but what race, no one knows. The history of this area of biological speculation appears as an introductory chapter in almost every textbook of vertebrate biology, and serves on the whole only to confuse the student and reduce to doubt his faith in the order of nature. Geoffroy Saint-Hilaire (1818) saw in the vertebrate merely a much-modified insect turned over on its back. Since his time, theory has succeeded theory: Hubrecht (1887) would have derived the vertebrates from the nemertean worms; Bateson (1884, 1885) from the acorn worms; Kovalevsky (1866) and Brooks (1893) from the tunicates; and Semper (1875), Minot (1897), Marshall (1879), Balfour (1881), Gegenbaur (1887), Eisig (1887), and Delsman (1913)—all great names in biology—tried by devious tricks of evolutionary legerdemain to obtain them from one or another type of annelid worm; Masterman (1898) from the jellyfish; while Gaskell (1896–1908) and Patten (1884–1912) confidently saw the prototype in the ancient arachnids, of which *Limulus* is a lonely survivor—thus returning the circle of speculation almost but not quite to Saint-Hilaire's inverted insects, though with little more cogency. Garstang (1894) was the first to look for the ancestors of the vertebrates in the larval rather than the adult state of some invertebrate, and to point to the larval echinoderms (starfish, sea urchins, sea cucumbers, and such) as a likely source; and Torsten Gislén (1930) and W. K. Gregory (1935) explored the larval echinoderm theory with encouraging results. The evidence is not to be found in the fossil record but rather in the comparative anat-

omy of living forms, including their larval stages, and
the most recent, and in many ways the most plausible
theory, is that of N. J. Berrill (1955), who would derive
the vertebrates from the free-swimming larval stage of
the pre-Cambrian tunicates.

As our present information stands, however, the gap
remains unbridged, and the best place to start the evolution of the vertebrates is in the imagination. The basic
features of the forerunner of the vertebrates—which we
will here call the 'protovertebrate'—were long ago laid
down in blueprint form by T. C. Chamberlain and require little modification (see Figure 4): it was an elongated, spindle-shaped, bilaterally symmetrical form that
possessed a stiffened and yet flexible 'backbone' for the
support of muscles so arranged as to produce powerful
lateral or eel-like movements of the body to propel it
through the water; the backbone and the skeletal muscles and the nerves to the latter, were divided into regularly repeated segments; the animal possessed a round,
jawless mouth by which it fed on plankton, infusoria,
and bottom-living plants; it had internal gills supported
by rudimentary gill arches, and sense organs and a brain
of a sort at the anterior end of the body, while the tail
was solid muscle and contained no viscera. We can only
guess at its size, but it was probably not more than one
or two inches long. Among the invertebrates no such
animal is known, for this blueprint resembles neither
jellyfish, sea anemone, flatworm, threadworm, earthworm, snail, oyster, squid, crab, lobster, *Limulus*, nor
insect, nor any of the larval stages of these forms. The
larval stage of the tunicates, as proposed by N. J. Berrill, presents a reasonable hypothesis, but nevertheless
if writers of science fiction want to import life from some
extragalactic star, they can do worse than to consider
the protovertebrate.

As we cannot say from what antecedents the vertebrates were evolved, neither can we with certainty name

the time of their evolution. Some have assigned this to
the Cambrian and some to the Ordovician period—the
former is accepted here. The Cambrian was ushered in
by the Charnian revolution, which in North America
took the form of what American geologists call 'the
Grand Canyon disturbance.' Where the Colorado River
has cut its gorge for 200 miles through Arizona, there
had been deposited a series of pre-Cambrian sedimentary rocks (the Grand Canyon System) composed of
conglomerates, limestones, shales, and quartzites, the
whole some 12,000 feet in thickness; during the Grand
Canyon disturbance these strata were uplifted and
broken by faults into mountains two miles high, mountains which were almost entirely reduced to sea level
again shortly after the orogenic disturbance passed.
Thereafter the Cambrian as a whole was one of continental submergence, and before its close more than 30
per cent of North America (to mention a well-studied
area) was covered by a sea that invaded the continent
in two great troughs, the Cordilleran trough extending
from Alaska to the Gulf of California and submerging
the area now represented by the Rocky Mountains, and
the Appalachian trough extending from Newfoundland
to the Gulf of Mexico and submerging the area roughly
corresponding to the later Appalachian Mountains. (Historical geologists are sometimes confusing by their
choice of names, which cannot simultaneously locate an
area in two dimensions of geography and the third dimension of time.)

The protovertebrate, with its spindle-shaped body
and segmentally arranged muscles adapted to rhythmic
contractions, is such, as Chamberlain suggested, as
would be evolved in response to the motion of rivers
flowing constantly in one direction, rather than in the
sea where local motion, if any, is a gentle ebb and flow.
No similar form has ever been evolved in the sea. It is
consonant with the evidence to suppose that the ances-

tors of the protovertebrate had radiated from the sea into the brackish estuaries bordering the continents when the Grand Canyon disturbance overtook them, and that continental elevation, by accelerating the motion of the rivers, fostered their evolution in the direction of the dynamic vertebrate form.

In wandering from the sea into the estuaries, the ancestors of the protovertebrate may have been driven by predatory enemies, or they may have been radiating into relatively unoccupied territory rich in algae and other food; what is more likely, they may have been just exploring the periphery of their world, seeking peace in the face of the never-ceasing drives of hunger and harassment. But peace they were destined never to find. That they immediately encountered trouble is revealed by the fact that when the fishes first make their appearance in the fossil record it is as heavily armored, bottom-living forms far removed from the hypothetical protovertebrate from which, by all evidences, they must have been derived.

Because of their heavy armor, these earliest fishes are collectively called the Ostracodermi (*ostrakon* = shell; *derma* = skin), and because they had no movable jaws they are grouped with a few greatly modified descendants (the living cyclostome fishes which include the lamprey, *Petromyzon*, and the hagfish, *Myxine*) in the class Agnatha, or jawless vertebrates.

By the Devonian period, when ostracoderm remains appear in the fossil record in fair abundance and a good state of preservation, they are represented by three major orders, of which type specimens are illustrated at the top of Figure 4. The Anaspida (*an* = without; *aspis* = shield), illustrated at the left by *Birkenia*, were very small fishes lacking any carapace but with the body entirely armored in heavy scales. The Cephalaspida (*kephale* = head; *aspis* = shield) (= Osteostraci), illustrated by *Cephalaspis*, are so named because of the heavy shield that covered the head. The Pteraspida (*pteron* = wing; *aspis* = shield) (= Heterostraci), illustrated at the right by *Anglaspis*, had a heavy shield over the whole front end of the body, and the trunk and tail were covered with large rhomboid scales.

Why all this armor of bony plates, scutes, and scales covering the animal from snout to tail? Some years ago A. S. Romer suggested that it was evolved as a defense against the eurypterids, the large scorpionlike creatures of the Silurian, shown at the bottom of Figure 4, because these presented the only visible enemies of the early vertebrates in their fresh-water habitat. The argument, however, is not wholly convincing, or at least is not the only argument that can be advanced. Though the largest American eurypterid, *Pterygotus buffaloensis* of western New York, grew to an estimated body length of seven feet (the largest arthropod of all time) and possessed pinching claws, the majority were but a few inches to a foot in length and had no pinching claws whatever. If not primarily mud crawlers like their descendant the horseshoe crab, the eurypterids, which had only heavy and clumsy paddles on the head, did not have the appearance of powerful swimmers. The mouth was a small opening on the underside of the head (as in the horseshoe crab), and bore toothlike processes beginning to function as chewing organs; it suggests the scavenging stage of an advanced clumsy mud-strainer rather than a 'predacious' and 'voracious' animal capable of driving the vertebrates into armored safety. Even though, as was probably the case, types like *Eurypterus remipes* struck their prey with their pointed tails as do the horseshoe crabs, and types like *Eusarcus scorpionis* injected poison as do their other offspring, the scorpions, their fearsomeness may be a matter more of psychologic association than of fact. Moreover, the eurypterids did not reach their climax until the Silurian; they are poorly represented in the Ordovician, and it may be questioned whether they were then or ever numerous enough to set the evolution of the entire vertebrate phylum in a pattern of external armor that was to persist for one hundred and fifty million years.

The eurypterid theory seems to the writer to be inadequate by itself to account for the armor of the ostracoderms. All samples of the fossil record from Silurian and Devonian times, from Spitsbergen to Colorado, suggest that some death-dealing enemy, swift, merciless and irresistible, lurked in every corner of the world. This enemy, we believe, was the medium in which the early vertebrates were undergoing evolution: it was an enemy they could not see but one that pursued them every minute of the day and night, one from which there was no escape though they deployed from Spitsbergen to Colorado—the physical-chemical danger inherent in their new environment: their fresh-water home.

Water diffuses rapidly through all physiological membranes in accordance with differences in osmotic pressure—more exactly, the diffusion pressure of water molecules—on either side. Salts lower the diffusion pressure of the water in which they are dissolved and consequently water diffuses from any dilute salt solution into a more concentrated one until the diffusion pressure is equalized. When the first migrant from the sea sought to invade fresh water, its blood and tissues were relatively rich in salts, a physical-chemical inheritance from its primordial salt-water home. (We may on straight extrapolation assume that at the opening of either Cambrian or Ordovician time the sea had at least one-half and perhaps five-sixths of its present salinity.) This saline heritage could not be wholly cast aside, and whenever the organism sought to migrate from salt into fresh water, water tended to move into the body by osmosis. Were this influx of water not arrested, the animal would have taken in so much water that it would have met death by dissolution as a gelatinous mass. Those mutant forms that possessed the advantage of waterproof armor survived to produce the ostracoderms, and, from the ostracoderms, all the higher vertebrates. If, incidentally, the waterproof armor served to protect them from the eurypterids, so much the better.

CHAPTER IV

THE KIDNEY

The fact that the theater of evolution of the early vertebrates was in fresh rather than salt water had momen-
tous consequences. In acquiring a heavy exoskeleton of
waterproof armor, the protovertebrate was transformed
from a dynamic, free-swimming form to a sluggish creature that perforce kept to the bottom of the rivers and
lakes where it groveled in the mud for food. However,
it came about that in time certain movable spines on
the armor of the ostracoderms evolved into the fins of
the Silurian and Devonian fishes, and from the pectoral
and pelvic fins there evolved the four legs of the tetrapods, permitting them to engage in amphibious life and
to establish the great empires of the reptiles, birds, and
mammals. Again, the armor of the ostracoderms required a complete reconstruction of the head and the
conversion of the primitive gill-arches into jaws; and
without jaws and the ostracoderm plates that persisted
as teeth, the jaw-bearing vertebrates would scarcely
have attained their predatory supremacy over the invertebrates. Without the predatory power of jaws and
teeth and the possibility of swift and accurate pursuit
of prey there would have been no evolution of the
distance-sense-organs of smell, sight, and hearing, of
elaborate muscular co-ordination, of prevision of how
to get from here to there and the possible consequences
of going from here to there—in short, there would have

been no centralization of the nervous system such as
ultimately produced the brain, and the earth would
never have known the phenomenon of consciousness, at
least of an order superior to that of the lobster, scorpion,
or butterfly.

But without an internal environment of relatively constant composition in which complicated nerves, muscles,
and glands could attain a high degree of elaboration
and function in security, it is unlikely that the fresh-
water fishes, with their elaborate sensory and motor
equipment, would ever have been evolved. It was the
evolution of the kidney that supplied the vertebrates
with this stable body fluid.

The most important constituents in the body fluid of all
vertebrates are water and sodium chloride—ordinary
table salt, the primary salt of the sea and of all plants
and animals. One cannot, in connection with living organisms, think of water without salt or salt without
water. All marine invertebrates are isomotic with their
environment—in other words, their body fluids have the
same salt content, and hence the same osmotic pressure,
as the sea water in which they live. The maintenance of
salt and water balance is therefore a relatively simple
matter. Excess salt is partly, if not entirely, excreted
through the respiratory epithelium; the problem of excreting water *per se* does not exist. The kidney is a glandular organ of greater or lesser complexity and highly
varied in structure in different animals; it is concerned
chiefly with the excretion of such waste products as cannot escape from the body by simple diffusion. It may be
assumed that such was the situation in the marine ancestor of the vertebrates. In the evolution of the vertebrates, however, the kidney came into a new and important role: it became the chief route for the excretion
of water from the body, an operation that had to be
carried out without loss of salt, and thus it came to be
responsible for the regulation of the composition of the

internal environment in respect to both water and salt, as well as for the excretion of waste products.

Water was available in excess to the protovertebrate, but the concentration of sodium chloride in this water was probably low and highly variable from river to river, from lake to lake, from rainy season to rainy season; and it is not too venturesome to think that the tenacious conservation of salt is one of the most primitive—if not the most primitive—of functions in the vertebrate kidney. The evolution of an equally tenacious conservation of water posed a problem to which entirely different solutions were to be found by the lungfishes, elasmobranchs, Amphibia, reptiles and birds (the marine reptiles and birds with unique patents of their own) and the marine teleosts; and yet another by the mammals—until after four hundred million years a small, nocturnal rodent can live in a desert burrow without water and the scientist can write in rhythmic heptads:

Salt and Water

In the beginning the abundance of the sea
Led to profligacy
The ascent through the brackish waters of the estuary
To the salt-poor lakes and ponds
Made immense demands
Upon the glands
Salt must be saved, water is free

In the never-ending struggle for security
Man's chiefest enemy,
According to the bard of Stratford on the Avon,
The banks were climbed and life established on dry land
Making the incredible demand
Upon another gland
That water, too, be saved.

Maurice B. Strauss, November 23, 1951.

Though the soft organs of the body are never preserved in the fossil record, the evolution of the kidney can be reconstructed with reasonable accuracy from data obtained from living forms. In this reconstruction we can begin with the premise that the body of the protovertebrate was divided into many similar, regularly repeated muscular segments. This segmentation persists in the skeletal muscles, nerves, and backbone of all the higher vertebrates. The viscera or hollow organs were, however, not segmentally divided but extended continuously from mouth to anus, and were contained in an unsegmented body cavity known as the coelom (*koiloma* = a hollow).

In the embryos of all vertebrates the viscera, on the one hand, and the segmented muscles and backbone, on the other, develop from different germinal layers—the viscera from the inner germinal layer or endoderm; the muscles and backbone from the middle germinal layer or mesoderm. The kidney develops not as one might expect from the endoderm with the viscera, but from the mesoderm (as also do the gonads and adrenal cortex), and, like the muscles and backbone, it starts out as a segmented structure. From this fact, and from the structure of the adult kidney in the primitive fishes, it is inferred that the kidney of the protovertebrate was also segmented, and that each segment of the body throughout the length of the coelom carried a bilateral pair of nephric (*nephros* = kidney) tubules. There is some evidence that the membrane lining the coelomic cavity primitively played a role in excretion, and it appears that the nephric tubules were first formed by multiplication of cells belonging to the coelomic membrane. On the interior of the body each of these nephric tubules communicated freely with the coelom by means of an open mouth or 'coelomostome' (*coelom + stoma* = mouth, also frequently called 'nephrostomes'); on the exterior they drained either through separate vents in the body wall like the scuppers of a ship, or into a common longitudinal groove in the skin. (See Figure 5a.)

These primitive nephric tubules probably also served to carry off the eggs and sperm, which were shed freely by the gonads into the coelomic cavity; embryonically, the gonads are derived from the coelomic membrane adjacent to the tissue which forms the nephric tubules, and this suggests that reproduction and excretion have had, from far back in vertebrate history, a close affiliation.

This segmented 'kidney'—if we may so designate a dozen or more pairs of primitive nephric tubules—is such as one might expect to find in a segmented animal in which the coelomic cavity participated in excretion and where a relatively feeble stream of coelomic fluid would serve to wash the excretion out of the tubules; and it may be accepted that it was with this meager equipment that the protovertebrate tried to enter the brackish lagoons or fresh-water rivers and lakes of the Paleozoic continents.

The armor of the ostracoderms served in its passive way to reduce the influx of water into the body to a minimum, but the gills and the membranes lining the mouth could not be insulated in this manner. Moreover, the ostracoderms must have swallowed considerable quantities of water when eating microscopic food, as do many living fresh-water fishes. Consequently the early fresh-water invaders had to compensate for this irreducible influx of water by increasing the rate of its excretion. It appears that no better way could be devised to get this water out of the body than to have the heart pump it out; and the simplest way to do this was to install, close to the open mouths of the pre-existing tubules, a filtering device in the form of a tuft of permeable capillaries (see Figure 5b). Under the hydrostatic pressure supplied to the arterial blood by the heart, water was filtered through these capillaries and drained out of the body through the tubules, while the blood cells and plasma proteins were retained in the capillaries and returned to the circulating blood. Such an elementary device persists today as the 'glomus' (*glomus* = ball) of capillaries which is formed in the first, transient kidney (the pronephros) in the embryos and larvae of nearly all fishes, as well as in the pericardial cavity of a few adult primitive forms (the hagfish, *Myxine*, and a few bony fishes).

Then in order to improve the drainage of the filtrate, the capillary tuft was inserted in the tubule, perhaps at first in association with an open coelomostome (see Figure 5c); ultimately, however, the tuft was pushed into the blind end of the tubule (Figure 5d) and the coelomostome disappeared. This is the structure of the *glomerulus* (diminutive of glomus) as it appears in the adult kidney of all the higher vertebrates. In the embryonic glomerulus the tubule is expanded into an enclosing sphere or capsule (Bowman's capsule) which serves to collect the filtrate and to direct it into the tubule without transit across the body cavity.

With the evolution of the glomerular nephron, there began a battle between the kidney and the reproductive organs that continued for three hundred and fifty million years. After the renal tubule, in taking on a predominantly excretory function, lost its opening into the body cavity (Figure 5c), it could no longer serve as a route of egress for sperm and eggs, and yet sperm and eggs had to get to the exterior if the organism was to reproduce. Moreover, as the body became encased in armor, the multiple, external openings of the segmentally arranged tubules had to be covered over and the kidney had to make for itself a new, internal conduit by which the urine could escape. This was accomplished by fusing together the tubules in the posterior segments on either side of the body cavity to form two ducts—the 'archinephric ducts'—into which the anterior tubules drained, these ducts emptying, with the intestine, into a common posterior vent or cloaca. But as the kidney built up the archinephric duct, the ovary and testis invaded it as a route of egress for their products, and thereafter the urinary system and the reproductive system became so entangled, especially in the male, that the two systems—excretory and reproductive—were not entirely separated until the reptiles were evolved—to the confusion of students of comparative anatomy, who sometimes have difficulty in allocating the proper anatomical and physiological priorities.

To sum up this confusion as succinctly as possible, the archinephric duct sometimes retains a pure urinary function (hagfish and lampreys); or it may carry both sperm and eggs through nearly all its length (Australian lungfish, sturgeon, gar pike, common frog and mud puppy); or the gonads may take over the anterior part of the duct, leaving the kidney only the posterior part or forcing it to develop a separate duct wholly independent of the archinephric duct—the pattern in the female even within the same species not necessarily conforming with that in the male (sharks and some salamanders); or the kidney may retain the archinephric duct entire while the testis develops its own conduit (bony fishes); or the kidney may abandon the fight in favor of the testis, when the archinephric duct becomes the spermatic duct, which carries the sperm from the testis to the seminal vesicles for storage; and in the last case the kidney has to build a new duct entirely of its own—the true ureter as it appears in the reptiles, birds, and mammals.

The ovary has on the whole been less invasive of the kidney's rights and, with few exceptions, the ova in all vertebrates are virtually thrown free into the body cavity to find their way to the exterior—that is, to the oviduct or uterus—as best they can, by means of open 'peritoneal funnels' (Müllerian ducts or Fallopian tubes).

The student of anatomy can find solace in the fact

that both urine and reproductive cells get out of the body one way or another in all extant forms—or they would not be extant, and perhaps also in the fact that this historic conflict between the kidney and the reproductive organs has had the most profound effects on moral values and psychopathology, though space prohibits an excursus into these topics.

In the higher animals the excretory system consists of two nearly identical kidneys located at the back of the abdominal cavity. In man each kidney contains nearly a million nephrons, each consisting of a glomerulus and tubule, these nephrons differing from each other only in minor and probably not significant details of structure. The nephrons drain into a treelike system of collecting ducts by which the urine is conveyed to the renal (the proper adjective pertaining to the kidney) pelvis and thence by way of the ureter to the bladder.

The embryologist Ernst Haeckel once said that 'ontogeny recapitulates phylogeny'—meaning that during their embryonic development organisms recapitulate their evolutionary history. Haeckel's aphorism must be accepted with many reservations, because during evolution patterns of embryonic development have been changed as much as the structure of adult organisms, but the aphorism is true in many important respects. And in several ways the evolutionary history of the kidney is recapitulated in its development. Most notably, in the embryos of all reptiles, birds, and mammals, two separate but abortive kidneys develop before the adult kidney is formed, a fact that recapitulates the long conflict between the testis and the renal tubules for the archinephric duct. The first embryonic kidney, or 'pronephros,' forms an archinephric duct by the fusion of a few of its tubules—just as this duct must have been formed in the ostracoderms—but in only a few of the lowest fishes do the pronephros and archinephric duct continue to function in the adult: in the hagfish

(*Myxine*), for example, the archinephric duct continues
to drain a single pair of nephrons in each segment of
the body, while the lamprey (*Petromyzon*) has simply
gone one step further by greatly multiplying the number
of nephrons draining into the archinephric duct. In the
higher fishes, the appropriation of the anterior part of
the archinephric duct by the testis has forced the kidney
to increase the number of nephrons attached to the posterior segments of the archinephric duct, thus producing
a more or less compact organ, the 'mesonephros.' While
in the reptiles, birds, and mammals, in which the kidney
has completely abandoned the archinephric duct to the
testis, the kidney must develop its own ureter. In the
mammals, both the pronephros and mesonephros are
only passing embryonic episodes, and the adult kidney,
or 'metanephros,' is formed by the local multiplication
of nephrons in the most posterior segments of the body
—as far away from the embryonic testis as it can get.

Biological patents carry no automatic expiration date
but yield only to new inventions that supersede them
because of greater effectiveness, and old and new may
long overlap in simultaneous operation or until the old
is put to some new use. Thus it was with the processes
of tubular reabsorption and excretion. It may be assumed that the coelomic tubules of the protovertebrate
reabsorbed some substances from the fluid draining
through them, in order to conserve them for the body;
but we can do little more than speculate about the nature of these reabsorptive operations. With the invention
of the glomerulus, however, it was necessary to speed
up all reabsorptive processes to a high rate, and probably to add new ones: because practically all the osmotically active constituents of the plasma had to pass, along
with water, through the semipermeable glomerular
membrane, since otherwise the limited pressure in the
glomerular capillaries would be insufficient to filter any
significant quantity of water. Thus increased tubular re-

absorption had to evolve in a parallel manner with increased filtration, until the kidney came to operate as an elaborate 'filtration-reabsorption' system working at top speed all the time.

It was through the evolution of this filtration-reabsorption system that the kidney came to be almost entirely responsible for the composition of the internal environment of the body, manufacturing it, as we have said, in reverse, by saving some things from the glomerular filtrate and rejecting others. Among the substances known to be reabsorbed by the tubules in man, for example, are water, sodium, potassium, calcium, chloride, bicarbonate, phosphate, sulfate, glucose, fructose, a large variety of amino acids, and several vitamins and hormones. There are certainly many others on which no quantitative observations are available.

Chief in this list, however, was water: by filtering large quantities of water and reabsorbing most of the filtered water (99 per cent in man), large quantities of waste products could be passively filtered through the glomeruli along with the water and then excreted in a urine that had been concentrated by the reabsorption of water. Although a fine balance had to be maintained between the reabsorption of salt and water in order to maintain the proper proportions of these substances in the body fluids, so long as the animal remained in fresh water or had an excess of water available to it, the excretion of waste products posed no problem other than the negative one of not reabsorbing them from the filtrate.

The filtration-reabsorption system served to excrete waste products not so much because it was primarily evolved for this purpose as because it automatically accomplished this result by letting some substances pass into the urine while saving others. It presents no paradox, therefore, that the kidney was long coming into its role as the sole excretory organ of the body: for example, in the fishes the chief nitrogenous end-product of protein metabolism, urea, is excreted almost entirely by simple diffusion through the respiratory epithelium of the gills. Salt, the most important inorganic constituent in the body fluid, is actively excreted by the gills of fishes in the interests of body-fluid regulation, despite the fact that the renal tubules are working steadily to reabsorb almost every trace of salt filtered through the glomeruli in order to save it from being lost from the body. In the adult Amphibia (which have no gills) the skin plays an almost equally important role in both salt and water balance. In short, full responsibility for body-fluid regulation did not devolve on the kidney until the first terrestrial animals took to living permanently on the land and the participation of both gills and skin in body-fluid regulation was wholly abandoned; only then did the filtration-reabsorption system come into its own as a highly efficient device serving both body-fluid regulation and the excretion of waste products.

And so it was with tubular excretion, a process by which the tubule cells remove certain substances from the plasma and deposit them directly in the tubular urine, thus supplementing the process of filtration. That tubular excretion is an invention dating back to the proto-vertebrate is plausible enough since specific secretory operations of this nature are common among the invertebrates. It is conceivable that a simple filtration-reabsorption system could have met all the requirements of the aquatic vertebrates—but, as it came about, the mechanism of tubular excretion was carried over into the glomerular nephron, and at a later day made possible the survival of three large groups, the marine fishes, the arid-living reptiles, and the birds, and it continues to contribute importantly to the excretion of waste products in man.

And so, also, with the kidney's blood supply, where old and new inventions overlapped and for a long time

functioned simultaneously. The simple coelomate tubule
had required for its reabsorptive and excretory operations a supply of blood, but since there was no need for
the blood to have a high pressure (as for filtration), it
was supplied in a random sort of way from miscellaneous veins draining organs in the posterior part of the
body—the tail muscles, ovaries, and others. (Figure 5a.)
This blood had already traversed the capillary bed of
these organs before it was distributed to the capillaries
of the renal tubules. The presentation of blood in this
manner to two capillary systems in succession is classi-
cally exhibited between the intestine and the liver,
where the nutriment absorbed into the capillary plexus
of the one is delivered directly to the capillaries of the
other; ever since the days of the anatomist, Galen, the
vein by which the blood makes this transit has been
called the 'portal vein' (from porta = gate). Anatomical
names are sometimes as conservative as organisms, and
the venous blood supply to the kidney came, by analogy,
to be called the 'renal-portal vein.'

When the glomeruli were evolved they had to be supplied with arterial blood at a high pressure in order to
effect filtration, but this blood was made to do a double
duty: after leaving the glomerulus it, too, was spread
out around the tubules in a network of capillaries where
it could sustain tubular reabsorption and excretion. (Figure 5b.) Thus the tubules acquired a double blood supply, the old one carrying venous blood, the new one,
arterial blood, but with no confusion: the capillaries of
the two systems simply fused with each other so that
it is impossible in any capillary to tell which blood is
which.

This double blood supply persists in the kidney of all
vertebrates below the mammals. In the latter the tubules
are supplied with blood only from the glomeruli (Figure
5d), and the renal-portal system appears only transiently
in the embryo as a flashback into history, and then (in

man) it degenerates into a small vascular network, the
'pampiniform plexus,' surrounding the spermatic duct—
itself a vestige of the archinephric duct long ago stolen
from the kidney by the testis. This circulatory plexus
now serves but to puzzle the unhistorically minded student of anatomy, and on occasion to undergo varicose
enlargement into the pathological condition known as
varicocele.

The microscopic structure of the renal tubule is much
the same in all vertebrates. This structure, as we know
its details today, throws little light on the many remarkable physiological operations which the tubule carries
out. In all forms below the mammals the tubule is differentiated into only two anatomically distinct segments,
a *proximal* (= near) segment adjoining the glomerulus,
and a *distal* (= remote) segment leading into the collecting ducts. In the lower vertebrates these two segments are connected by a short, narrow intermediate
segment which carries whiplike hairs or cilia that, by
beating downstream, help propel the urine from one segment to the other.

By a variety of techniques modern renal physiology
has greatly advanced our knowledge of the function of
these segments in man as well as in experimental animals. It is known that the proximal segment reabsorbs
many of the valuable constituents of the glomerular filtrate, perhaps nearly all of those substances which we
have enumerated earlier as being conserved by the kidney. In the mammals, the proximal segment also reabsorbs approximately seven-eighths of the salt and water
of the filtrate, and in this operation unreabsorbed waste
products in the tubular urine are concentrated eight-
fold.

The proximal segment is also responsible for the excretion of a large variety of both naturally occurring and
foreign substances. The reason why some substances are
excreted by the tubule and others are not is poorly understood, but it appears that this process helps to rid the

body of certain types of substances which cannot be utilized and the accumulation of which in the blood would be injurious. It is remarkable, however, that many synthetic substances which the organism has never encountered in its evolutionary history are excreted just as efficiently as are substances formed naturally in the body, probably because of the presence of certain common, chemically reactive groups in the molecule.

In the distal segment the greater part of the remaining sodium is reabsorbed, leaving the corresponding quantity of water free for excretion as what we may call osmotically free water since it contains no salt, thus fulfilling the function of the fresh-water kidney by excreting water without loss of salt. Also, in the distal segment, the urine is acidified and ammonia is added in exchange for sodium and potassium, operations that permit the excretion of acids without loss of these important salts, and that conserve for the body its 'alkali reserve' (sodium bicarbonate) and maintain constant the hydrogen ion concentration of the blood. So important is this steady state that a very minute increase in the hydrogen ion concentration of the blood induces the clinical state of coma by interfering with the function of the brain. It is probably also by the distal segment that the excretion of potassium, next in importance after sodium and the hydrogen ion in the over-all composition of the body fluids, is controlled.

As might be anticipated, natural metabolic end-products (urea, creatinine, creatine, uric acid, and others), as well as foreign substances, are handled by the renal tubules in different animals in a variety of ways, and the only generalization possible is that, in the formation of urine in every glomerular animal, both tubular reabsorption and excretion play a part.

This generalization poses one of the major problems of renal physiology: what substances are excreted by filtration only, and what substances by filtration supplemented by either tubular reabsorption or tubular excretion? The answers to this question can be obtained only by measuring the initial process of filtration: that is, by determining the volume of water filtered per unit time in any individual under a given set of circumstances. (This is, of course, far greater than the rate of excretion of water as urine because most of the filtered water is reabsorbed by the tubules.) Given a method of determining the rate of filtration of water, then simultaneous measurements of the rate of excretion of any other substance permits one to determine whether it is reabsorbed from this filtrate, or, alternately, added to the filtrate by tubular excretion and, in either instance, precisely to what extent. A hundred years and more elapsed in the study of renal physiology before this method became available, and indeed before the problem was even phrased in this simple, quantitative way. When the answer came, it came not primarily from studies of renal function in man but from studies in the comparative physiology of the kidney in a large variety of animals.

The development of renal physiology as a science might be said to have started in ancient times, since the physicians of Egypt, Babylonia, Greece, and Rome had acquired certain elementary notions of cause and effect in such sequences as the obvious increase in urine flow (diuresis) that results from the ingestion of large quantities of water, the decrease in flow (oliguria) that results from water deprivation, and the discoloration or odor of the urine that follows the ingestion of certain foods. However, the physicians of ancient times had no knowledge of the fine anatomy of the kidney and less knowledge of physics and chemistry, and consequently their observations lacked any proper interpretation.

Modern renal physiology (like all physiology) begins properly with William Harvey's demonstration, in 1628, that the heart is a pump that keeps the blood in steady circulation around the body through the arteries and veins. In establishing the reality of this circulation, Harvey utilized observations on many animals—scarcely a creature native to the English countryside escaped his investigation—and each contributed to the affirmation of the thesis that broke the fourteen-century-old authority of Galen and laid the foundations for modern physiology and medicine.

It was shortly after Harvey's death that Marcello Malpighi (1666), for the first time systematically exploring the fine structure of the organs of the body with a microscope, discovered the glomeruli of the kidney—they are still referred to in many texts as 'Malpighian corpuscles'—and incorrectly described them as 'glands,' failing to see the minute blood vessels or capillaries which he himself had just discovered in the lungs of frogs. Four years previously Lorenzo Bellini had traced backward from the renal pelvis the larger collecting ducts of the kidneys, to see them break up into finer and finer branches as dissection proceeded into the interior of the organ, and demonstrated that these were not solid fibrous strands but hollow tubes—today still referred to as 'Bellini's ducts.' Malpighi surmised, but did not prove, that his corpuscles were connected with Bellini's ducts and that they played an important part in urine formation.

Here the matter rested for almost two centuries, until in 1842 William Bowman, using microscope equipment not available in Malpighi's time, demonstrated the true relations of the capillary tuft to the tubule: that in each 'Malpighian corpuscle' the crescent-shaped space around the capillary tuft drains freely into the lumen of the tubule. This space with its investing membranes is today called 'Bowman's capsule.' Bowman, like Harvey and Malpighi, studied every type of animal available to him: his classic paper describes the 'corpuscles' and 'tubes' as they appear in the badger, dog, lion, cat, mouse, squirrel, guinea pig, horse, parrot, tortoise, boa, frog, and common eel, as well as in man.

It remained for the physiologist Carl Ludwig, in 1842,

to adduce the first evidence that the 'corpuscle' functions in the wholly passive manner of a minute filter, and to propose that the filtrate thus formed is sufficient in quantity to carry into the 'uriniferous tubes' all the waste products, which are finally concentrated by the reabsorption of water during the passage of the filtrate down these tubes.

Early in the present century Ludwig's filtration theory received strong support from E. H. Starling, to whom physiology is indebted for many basic principles pertaining to the circulation, and a little later from A. R. Cushny, who in 1917 published the first definitive work on urine formation, which (though not so entitled) is familiarly known to all students as the 'Modern Theory of Urine Formation'—the word 'modern' here meaning that, with Ludwig and Starling, Cushny visualized this process in modern terms of physical chemistry, free from the ancient vitalism that still lingered in renal physiology. Between 1924 and 1938, the Ludwig-Cushny theory of glomerular filtration was removed from the realm of speculation to the category of demonstrated fact by A. N. Richards, A. M. Walker, Jean Oliver, and their collaborators, who in a brilliant series of studies on a variety of animals collected the minute quantity of fluid available in Bowman's capsule and at various points in the tubule by means of a micro-pipette, subjected it to precise, qualitative analysis, and showed that this fluid did in fact have just such a composition as was required by Ludwig's filtration theory.

The definitive demonstration of tubular excretion was afforded in the years between 1923 and 1934 by E. K. Marshall, Jr., and his collaborators, first in frogs, rabbits, and dogs, and then in a theory-shaking series of observations, made simultaneously with those of J. G. Edwards, on the aglomerular marine fishes in which urine formation is entirely dependent on tubular excretion.

From 1930 on, renal physiology has been chiefly con-

cerned with quantitative studies of the filtration rate and
the renal blood flow and the physiological regulation of
these functions, and of the detailed operations involved
in tubular reabsorption and excretion. Students who
have been away from their elementary textbooks scarcely
ten years deplore the fact that the subject matter cur-
rently appearing in the technical journals is decorated
with mathematical equations, intricately concerned with
enzyme systems and energy transformations, and casu-
ally displays utterly new concepts developed since
Cushny's time—all welded together by quantitative principles. This current literature is, of course, deeply concerned with man, with his heart and circulation and
kidneys, and his misfortunes in disease, but human
physiology is only a part of the story: even as the foundations of renal physiology were established through
studies on many diverse animals, so today's research
ranges profitably through the spectrum of sharks, rays,
skates, the fresh- and salt-water fishes, the frog and mud
puppy, the alligator and the snake, the chicken, and a
variety of mammals. The dog has contributed more to
our knowledge of renal function (as it has to many other
areas of medical science) than any other animal not excluding man himself; not only is it docile, intelligent, and
and easily trained to co-operate, but also, in respect to
renal function, it closely resembles man. A dog we called
Blitz, who came into the writer's care in October of
1939 and had to be sacrificed in 1951 because of senility,
served for twelve years as the co-operative subject of
renal research for many young investigators whose
names are now illustrious in the annals of American
medicine, as well as for students from Italy, France, Hol-
land, England, Denmark, Sweden, Argentina, Chile, and
Mexico. It was largely on Blitz that the writer and his
collaborators worked out the method for measuring the
renal blood flow that is now used to measure that function in man in every country named.

The introduction of a reliable method for measuring the filtration rate in any animal at any time marks the beginning of what may be called the quantitative era in renal physiology. Given this datum, simultaneous measurement of the rate at which any other substance is excreted makes it possible to determine whether it is reabsorbed or excreted by the tubules, and to precisely what extent.

For measuring the filtration rate a physiologically inert substance is required that (a) is completely filterable through the glomeruli, (b) is neither synthetized nor destroyed by the tubules, and (c) undergoes neither addition nor subtraction by the tubules as it passes from the glomeruli to the renal pelvis. The polysaccharide inulin —a soluble, starchlike substance obtained from dahlia roots, the Jerusalem artichoke, chicory, and other vegetable sources—is generally accepted as best fulfilling these specifications. Inulin is nontoxic when properly prepared and adequate doses can be administered to animals and man with ease, while abundant evidence indicates that it undergoes neither tubular reabsorption nor excretion in any species.

Inulin must be introduced into the plasma by an appropriate method (generally by intravenous infusion) and at such a rate that the concentration in the plasma will be maintained at a suitable level throughout the period of observation. Urine is then collected accurately at short intervals, usually ten to fifteen minutes in mammals, at much longer intervals in cold-blooded forms. The rate of excretion of inulin (expressed in milligrams per minute), divided by the quantity in each cubic centimeter of plasma, gives directly the volume of plasma filtered per minute—since for each milligram of inulin excreted a corresponding quantity of plasma must be filtered in order to transfer the inulin into the urine.

This procedure is what is known in renal physiology as the 'clearance method,' since it consists of the determination of the volume of plasma which is 'cleared' of a given substance per unit time. If, as in the case of inulin, the substance is 'cleared' from the plasma only by filtration, with no tubular reabsorption or excretion, the rate of plasma clearance (expressed as cc. per minute) is equal to the filtration rate. Where a substance is filtered but reabsorbed by the tubules, the rate of clearance will be less than the filtration rate; or alternatively, where a substance is excreted by the tubules in addition to being filtered, the rate of clearance will be greater than the filtration rate.

Where tubular excretion is present, the rate of clearance may be so large that it affords a measure of the renal plasma flow—which is the upper limit of any clearance value, since no more of any substance (not synthesized by the kidney) can be excreted into the urine than is brought to the kidneys in the blood plasma per unit time. In principle, the measurement of the renal plasma flow is as simple as the measurement of the filtration rate: given a substance that is readily excreted by the tubules (as well as by the glomeruli), all (or nearly all) of the substance contained in the arterial plasma entering the kidney may be removed in a single circulation and concurrently deposited in the urine. If one knows the quantity contained in each cc. of plasma and the concurrent rate of excretion (in mg. per minute), division of the second figure by the first gives directly the renal plasma flow, and appropriate correction for the hematocrit (or volume of cells in the plasma) gives the renal blood flow. Thus it is possible with selected substances to measure the renal blood flow in man and experimental animals without any disturbance other than an intravenous infusion, removal of blood samples from a vein, and the accurate collection of urine by means of a catheter.

The substance now most widely used for measuring the renal blood flow in man and other animals is p-aminohippuric acid, commonly referred to as PAH. In normal man, for example, 91 per cent of the PAH contained in the renal arterial plasma is removed and excreted into the urine in a single passage of the blood through the kidneys. (That this figure is not 100 per cent is partly attributable to the fact that some blood passes from the renal artery to the renal vein by way of non-excretory channels in the renal capsule, fat and other adventitious tissue.) The 'extraction ratio' of PAH has been accurately determined many times in man by the simultaneous analysis of arterial and renal venous blood, the latter collected by a long, flexible catheter passed into the renal vein by way of an arm vein and thence down the vena cava. This procedure must be carried out under strict surgical conditions, but in the hands of experienced physicians presents negligible dangers.

The collection of urine and blood in fishes and other cold-blooded animals offers little more difficulty than in man. Most fishes, other than the elasmobranchs, and all the Amphibia have a urinary bladder of sorts, sometimes quite large. In such forms the bladder can be emptied by catheterization and the urinary papilla closed with a purse-string ligature; or a retention catheter to which a small rubber bag is attached can be fastened in the papilla and the bag emptied at convenient intervals, a technique first used in fishes by E. Herter in 1891. Where no urinary bladder is present, as in the elasmobranchs, urine can be collected from the urogenital sinus by retention catheter and rubber bag. Blood can be collected in a hypodermic syringe without injuring the fish by puncturing the dorsal aorta from the ventral side of the tail while the fish is held in a V-shaped trough at an angle of 45°, with the fish's head under water. With practice, all operations can be completed in a few minutes and with no serious physiological disturbance to the animal.

Quantitative observations on the filtration rate and the renal blood flow, as well as on detailed tubular transport mechanisms in many types of experimental animals and in man, now afford a body of precise knowledge on the kidney not matched by that available on any other organ of the body. At one pole these observations are of interest in relation to the regulation of renal function in man and the disturbances in this function caused by disease; at the other pole interest is focused on the nature of the physiological mechanisms by which renal function is controlled, and on the biochemical nature of the cellular mechanisms by which various tubular operations are carried out.

CHAPTER V

THE ELASMOBRANCHS

Simple as the primitive glomerular-tubular kidney was,
it has served the fresh-water fishes for some five hundred
million years, and it appears to have served the ostracoderms equally well, since they increased in variety and
numbers in the fresh-water rivers and lakes of the Silurian and Devonian periods. We have noted that by the
Devonian, the ostracoderms had become diversified
into at least three major orders: the Anaspida, the
Cephalaspida, and the Pteraspida. Whether any of these
now extinct forms occasionally invaded the sea we do not
know, but in view of the fact that their degenerate
descendants—the lamprey, *Petromyzon* (thought to be
a derivative of the cephalaspid-anaspid root), and the
hagfish, *Myxine* (thought to be a derivative of the
pteraspids)—can live in salt water, it seems likely that
they may have done so, but that breeding or other habits
prevented them from establishing themselves permanently in this habitat.

It has been emphasized that evolution has no maintained direction, but rather many directions sustained
only by the opportunistic calculus of probabilities. In the
ostracoderms random possibility took some new directions that were ultimately to prove of great value to the
vertebrates. In the cephalaspids the primitive method
of laying down the waterproof covering had been trans-

formed by the appearance of bone cells (osteoblasts) which crawled into the irregular spaces of the armor and, in the walls of blood and lymph vessels, began to build true bone of calcium phosphate—a structure here making its appearance for the first time in the animal kingdom. Other bone-destroying cells (osteoclasts) followed them and continually dissolved this bony deposit almost as fast as it was laid down, so that bone could be formed anew and the animal could grow. Internal layers or rods of bone came to be pitted against external plates and scales until this new bony tissue, the precursor of the internal skeleton of the higher animals, became the strongest structural material in the animal kingdom. Because bone could not be used in the embryo, which must grow and constantly reconstitute its anatomy, its place was taken during the embryonic period by soft cartilage, an ostracoderm invention that continues in the higher vertebrates to connect bone to bone and bone to muscle even in the adult animal.

The cephalaspids also began the transformation of the armor around the head into a skull with cavities for the organs of smell, vision, and hearing, and with a trough to accommodate the rapidly enlarging brain; they invented the shoulder girdle by which the thrust of muscles could be transmitted to the head in order to steer it and push it into the mud in search of food; and, in feeding or in seeking shelter from their enemies among the lights and shadows, they were guided by a light-sensitive eye in the top of the head—the so-called 'pineal eye,' which is believed to be the precursor of the pineal gland in the higher vertebrates—which looked straight upward and was probably connected by nerve fibers with the pituitary gland.

The ostracoderms possessed no fins, but in many forms the armor had developed large spinous processes, which, when located in points of advantage, afforded the animal a fulcrum for anchorage. The cephalaspids and pteraspids had single dorsal spines, the cephalaspids and anaspids had a pair of pectoral spines, all of which probably served as balancing and anchoring organs. Though possibly of limited use for locomotion, these spinous processes, developed as outgrowths of the heavy body armor, were to supply the nodal points for the evolution of the fins of the later fishes.

None of the ostracoderms had movable jaws articulated with the skull, for which reason they are placed with the living cyclostomes (*kyklos* = circle; *stoma* = mouth) in the superclass Agnatha, or jawless vertebrates. The mouth, as in the protovertebrate, consisted of a round opening with hairlike appendages adapted for straining microscopic animals and plants from off the bottom of the streams. However, the armor around the mouth necessarily took the form of movable plates, and these plates afforded the raw materials for the later evolution of a movable lower jaw hinged to the cranium, a notable event because it set the pattern for the evolution of all the later jaw-bearing vertebrates, collectively known as the Gnathostomata (*gnathos* = jaw; *stoma* = mouth).

The Silurian saw the transition from the bottom-living, filter-mouthed ostracoderms to the free-swimming, jaw-bearing fishes, and by the Devonian the vertebrates had acquired the form of active, predatory fishes, the Placodermi (*plax* = tablet or flat plate; *derma* = skin)—incidentally the only one of the eight classes known to vertebrate history which has become extinct. By the Middle Devonian these placoderms were represented by several orders (Figure 6).

The Acanthodii (*akantha* = spine or thorn; *odiosus* = hateful), or 'spiny sharks,' were covered with small, flat, bony scales. They had one or two large dorsal spines and, on the ventral side of the body, two symmetrical rows of spines running from the pectoral region to the anus and ranging in number from two to seven pairs. Though these spines were probably not movable they carried attached to the rear edge a web of skin which made them more effective as stabilizers. All the higher fishes are presumed to be derived from a primitive acanthodian root.

The Arthrodira (*arthron* = joint; *deire* = neck), commonest of Devonian vertebrates, were large fishes derived from some unidentified acanthodian. They possessed a heavily armored head flexibly joined to the armored body by a ball-and-socket joint, so that for the first time the animal could raise and lower its head as well as move it from side to side, and to implement this motion they had elaborated on the shoulder girdle, to which the muscles that moved the head were anchored. Some (as *Coccosteus*) possessed paired pectoral and pelvic fins, the former still carrying the old ostracoderm spines. The posterior part of the body generally lacked armor, indicating that improvements in the structure of the skin had enabled them to begin the abandonment of heavy waterproofing.

The Antiarchi (*anti* = opposite; *archos* = anus) were grotesque little fishes related to the arthrodires. They possessed jointed flippers or creepers that were neither spines nor fins and resembled nothing before or since, but which are accepted as having been derived from spines. It has been suggested that, peculiar in all other ways, they also possessed lungs—certainly, if true, a case of 'convergent' evolution, since they are in no way related to the later air-breathing fishes.

The Stegoselachii (*stegos* = roof; *selachos* = the Greek name for fishes having cartilage instead of bones) lacked the ball-and-socket joint in the neck but possessed pectoral fins attached directly to the shoulder girdle, and in the brain and gill arches foreshadowed the later elasmobranch fishes.

Whereas spinous processes were once thought to be a sign of extreme evolutionary specialization, the spines of the placoderms are now seen to be derived from the ostracoderm armor and to have served to give the animal

anchorage in feeding, to stabilize it in swimming, and
perhaps to protect it against voracious enemies. These
fishes now all possessed powerful jaws with which to
seize and devour prey; most of them had a propulsive
tail; most of them had improved on the pectoral girdle, to
which ultimately were to be attached the muscles that
moved the head and pectoral spines or fins; and most
of them had developed a large and elaborate brain case,
this master nerve ganglion needing room for enlargement
as new sensory and motor apparatus was evolved.

Whether the origin of the ostracoderms is placed in the
Cambrian or in the Ordovician period, a long time
elapsed before any of their progeny permanently invaded the sea—at a minimum, half of the Ordovician and
all of the Silurian, or approximately ninety million years.
Since both animal and plant food must have been much
more abundant in the sea than in the fresh-water lakes
and rivers, this long delay suggests that insurmountable
physiological limitations rather than competition with
the invertebrates held them back from migration. There
is nothing about either the Cambrian or Ordovician to
afford an alternative explanation. Like the Cambrian, the
early Ordovician was characterized by widespread continental submergence, fully half of the present North
American Continent being covered by seas in which giant
cephalopods—ancestral to the pearly nautilus, cuttlefish,
squid, and octopus—and almost equally large trilobites
held the dominant position. The Ordovician saw the appearance of the first true corals, starfishes, and clams,
and the wide diversification of other invertebrates. Yet
the only remains of the ostracoderms are a few famous
'first fragments' which have been recovered from Ordovician fresh-water deposits near Cañon City and related
strata in the Big Horn Mountains and the Black Hills.
The period closed with the Taconic revolution, which
raised mountains of that name in the area from Newfoundland to New Jersey; the highest of these peaks had

disappeared before half the next geologic period had elapsed, but their eroded roots along the St. Lawrence and Hudson River valleys show the Ordovician strata turned almost on end.

After the Taconic revolution the seas again spread across the continents in the Silurian period, and a warm climate carried coral reefs to the Arctic Circle, while invertebrates were prevalent in all parts of the world where fossil beds are found. The most distinctive newcomers among the invertebrates were the eurypterids or 'sea scorpions,' which had been derived from the Ordovician trilobites; although confined to a few limited horizons of a fresh-water origin, the eurypterids are common fossils where they do occur. Land plants also made their first appearance, and possibly the first air-breathing animals, the scorpions, derived from a primitive eurypterid stem. It is interesting that these should have made the advance from aquatic to terrestrial life at least one geologic period if not two periods ahead of the vertebrates, but these terrestrial invertebrates were much smaller animals and continued to breathe by means of modified gills. In the abundant marine deposits of the Silurian, a period forty million years in length, invertebrate evolution was going on apace; but there is still no trace of marine vertebrates.

Though it is not invariable, the pressure of natural selection is frequently of such intensity as to seem to operate not through its subtler modes but by the very threat of death, by calling forth new adaptations as the only alternative to sudden and complete extinction. And it appears to have been such a catastrophe-impending climax that finally forced the vertebrates to seek refuge in the sea. At the close of the Silurian the earth began to heave again in another disturbance that raised a range of mountains higher than the present Alps, and extending in a great curve four thousand miles long from the north of Greenland eastward through Spitsbergen, south through Norway, and westward into Scotland and northern Ireland. Throughout Norway and Sweden the older Cambrian and Silurian formations were folded, over-turned, and overthrust along individual fault planes for distances as great as twenty to forty miles. Of this moun-tain system, one of the greatest the world has ever seen, the low and rounded Caledonian Mountains that com-prise the Scottish Highlands are all that now remain. Another range stretched across France, Germany, and Austria, while still others were formed in northern Africa and the Irkutsk basin of Siberia.

The Devonian period, which followed this Caledonian revolution, is identified for all geologists with what in the British Isles has long been called the 'Old Red Sand-stone,' a geologic formation underlying the so-called 'Coal Measures' of the Carboniferous. In most parts of the world the Devonian strata are of considerable depth, and in England, where the geologist Sir Roderick Murchison first studied the Old Red in Cornwall and Devonshire, and renamed the system for the latter, they reach a thickness of 10,000 to 12,000 feet; while in Aus-tralia the sedimentary and volcanic strata exceed 30,000 feet in depth and represent the most severe disturbance that continent has ever experienced.

For the marine invertebrates, the Devonian was a heyday of evolution. Corals were building giant islands in the warm seas, which extended at times up to the Arctic Circle, and great bivalves, the forerunners of mod-ern clams and oysters, were in their ascendancy and com-peting with snails, starfishes, sea urchins, squid, and cut-tlefish. For the continental fishes, however, the Devonian climate proved to be a mixed blessing. The Old Red of Europe, as well as the Devonian strata elsewhere, is primarily of continental origin and consists of conglom-erates containing great river-eroded stones, sandstones formed in the channels of streams, and siltstones and shales originally deposited as mud in quiet waters and frequently marked with cracks where successive layers were dried by exposure in the alternation of extreme wet and dry seasons. The red sandstone that gave the system its earlier name is an oxidized iron-containing rock that must have been formed from mud periodically exposed to air; but the thick layers of greenish-gray sandstone, siltstone, and gray shale interlarded with the red sandstone must have been formed in more permanent shallow waters, or in perpetually moist mud lakes. Whereas in the Lower Devonian the continental waters had been displaced by mountains drained by torrential rivers, by the Middle Devonian, when these mountains had been eroded to a considerable extent, large areas of the lowlands had become deserts of wind-blown sand and all habitable portions of the continents were subjected to extremes of climate. As annual drought succeeded annual flood, the evanescent lakes gave way to stagnant pools and hard mud flats in a cycle such as is commonly seen today in Australia, Central Africa, Eurasia, and the western part of North America; and most of the Devonian, fifty million years in length, presents a geologic record of, in alternation, too much and too little rain. The vertebrates, now represented by fishes of an advanced type, had to choose between living in stagnant pools and dry mud flats for half the year, or seeking sanctuary in the stable waters of the sea.

Two primitive groups of fishes chose the latter course. One, the Arthrodira, became extinct at the end of the Devonian and we cannot even speculate how they maintained their salt and water balance, or why they became extinct. But the second group survived to establish the marine cartilaginous fishes, the class Elasmobranchii (= Chondrichthyes), represented today by the sharks, rays, skates, and chimaeroids.

It is appropriate to emphasize here that there are two great groups of fishes that differ from each other in many notable respects: the cartilaginous fishes or the Elasmobranchii (*elasmos* = plate; *branchia* = gill), or, as they are sometimes called, the Chondrichthyes (*chondros* = cartilage; *ichthys* = fish), which in respect to evolution are very primitive and very old; and the more recent and so-called 'higher' fishes, the Osteichthyes (*osteon* = bone; *ichthys* = fish), which include all the other and so-called 'bony' fishes of the seas and fresh waters of the world. It is only with the cartilaginous fishes or elasmo-branchs that we are immediately concerned.

When the ancestors of the elasmobranchs were driven by the widespread aridity of the Devonian to seek sanc-tuary in the sea, they faced (as did the later bony fishes) a major physiological problem. The sea contained a relatively high concentration of salt (even as it did back in the Cambrian when the protovertebrate had abandoned it to take up residence in the continental fresh waters), and with the invasion of this salt water, the osmotic relations between the organism and its ex-ternal environment were completely reversed. Where once the difficulty had been the influx of excess water into the body, now the difficulty lay in excessive loss; the greater salt content in the sea water, by lowering the diffusion pressure of water below that of the blood, caused water to move out of the body through the gills and oral membranes (if not through the skin, which now, thanks to the impervious scales that had been evolved from the armor of the ostracoderms, was a fairly water-proof structure). Without provision to arrest or offset this constant osmotic loss of water, the animal would die of exsiccation just as certainly as though it were being slowly dried in the air and sun.

In theory, a fish moving from fresh water into salt water might solve its water-balance problem by drinking sea water and excreting the salt in a urine osmotically more concentrated than the sea, thereby maintaining it-self in salt and water balance. Natural selection, how-ever, does not work along the lines of theoretical blue-prints but on the raw materials randomly supplied by mutation and, as evolution worked out, the capacity to make a highly concentrated urine—or even one more concentrated than the blood—was not to be achieved in any significant degree until the evolution of the mammals. For this invention, three hundred-odd million years in the future, the Devonian fishes could not wait.

So when the elasmobranchs were driven from fresh water into the sea they sought another solution, and found one so simple that it would probably escape the imagination of the most teleologically minded biologist. They simply reduced the renal excretion of urea, letting this substance accumulate in the blood and tissues until it reached concentrations of 2.0 to 2.5 per cent (figures to be compared with 0.01 to 0.03 per cent in all other vertebrates). As this urea accumulates it contributes its share to lowering the diffusion pressure of water in the blood, until the latter falls below that of sea water and water begins to move by passive osmotic absorption into the body through the gills. The end result is that the animal, instead of losing water to the sea water, draws water out of sea water at no direct physiological expense.

For reasons probably not connected with the retention of urea, the skeleton of the cartilaginous fishes, as the name reveals, is not calcified into true bone. The general tendency in evolution is to lose bone rather than to gain it, and in the elasmobranchs this loss has gone to completion: they do not have a calcified bone in the body and are therefore easily dissected with scalpel and small scissors, for which reason the common dogfish is universally used for dissection in the teaching of comparative anatomy. The skin is generally covered with horny scales or dermal denticles (shagreen) having a structure very similar to that of teeth: an inner pulpy core surrounded by a layer of calcareous dentine and, outermost, a film of enamel secreted by the underlying ectoderm. This skin is impermeable to both water and urea, and represents an effective, pliable armor.

The term 'habitus,' borrowed from botany, is used by the paleontologist to designate any adaptive feature in

an organism that 'fits' it to its 'habitat,' and it is convenient to speak of the 'urea-retention habitus' of the
elasmobranchs, since it is such a unique adaptation to
salt-water life. It is one of the most strikingly simple
means known for automatically maintaining an important homeostatic state. This habitus is characteristic of
all the living cartilaginous fishes—the sharks, rays, skates,
and chimaeroids—but of no other living forms. Had this
fact been known when the biologist named this great
group of fishes it could well have supplied him with a
better name for the class as a whole. We shall not undertake the responsibility of renaming the Elasmobranchii; but if and when anyone does, it may be emphasized that it is *urea* which is conserved, and not
urine. It is clear that for three hundred and fifty million
years these fishes have had a wholly adequate renal function and that they are not 'sick' fishes suffering from
renal degeneration, as was suggested by early investigators in this field.

In other fishes urea is entirely excreted by the gills,
and as a step in the evolution of the urea-retention
habitus it was required that the elasmobranchs decrease
the permeability of the respiratory epithelium of the gills
to such a degree as to reduce the outward diffusion of
urea to a minimum. Urea is one of the most diffusible
substances known and, in the elasmobranchs as in all
other animals, it is distributed fairly uniformly throughout the body water. We know almost nothing about the
properties of cells that determine such molecular features
as permeability to various substances, and all that can
be said is that the respiratory epithelium in the elasmobranchs holds upwards of 2.0 per cent urea in the blood
against a zero concentration in sea water, and this without seriously impairing the permeability of the epithelium to oxygen and carbon dioxide, which must be continuously exchanged between the blood and sea water
for respiratory purposes. In the mammals the renal tu-

bules are almost equally impermeable to urea and maintain a high concentration in the urine as against a low
concentration in the blood, although they are simultaneously carrying on a variety of chemical exchanges between these two fluids. But these two cellular structures
—the respiratory epithelium of the elasmobranchs and
the renal tubules of the mammals—are the only physiological membranes known that are virtually impermeable to this substance.

The second step in the evolution of the urea-retention
habitus was to recover the urea from the glomerular filtrate by tubular reabsorption, as glucose and other valuable substances had been reabsorbed by the earlier vertebrates. How the elasmobranchs do this is also a mystery;
the reabsorptive process is an active one specifically involving urea molecules, and it is the only instance of
active reabsorption of urea known. It appears that this
reabsorptive operation is carried out not by a unique segment but by tubule cells which in other animals do not
have this function. Whatever the mechanism of reabsorption, it saves some 90 per cent of the filtered urea
from excretion.

The elasmobranch kidney maintains such a nice balance between the filtration of urea and its conservation
by the renal tubules that, with allowance for a small
and unavoidable loss of urea through the gills, the osmotic pressure of the blood is maintained at a level that
will draw just sufficient water from sea water to meet
the animal's needs. The mechanism appears to work with
great simplicity: as the water absorbed through the gills
dilutes the blood, the urine flow increases, increasing the
excretion of urea and thereby reducing the blood urea
concentration, which in turn reduces the absorption of
water through the gills, reduces the urine flow, and starts
a new cycle of urea retention, so that the animal is always supplied with enough free water to meet its urinary
requirements.

This description can happily be supplemented by reference to what may be called the rectal 'salt' gland of the dogfish (analogous to the nasal 'salt' of the marine birds and reptiles described in Chapter XI). All the elasmobranchs, apparently, possess a glandular appendage to the gut, located in the dorsal mesentery and draining into the posterior gut behind the spiral valve.

The function of this gland remained unknown until the summer of 1959, when Dr. J. Wendell Burger, working at the Mount Desert Island Biological Laboratory in Salisbury Cove, Maine, demonstrated that in the spiny dogfish, *Squalus acanthias*, the gland secretes a colorless, neutral sodium chloride solution. This secretion is isosmotic with the plasma but contains nearly twice the sodium chloride concentration of the latter (ca. 500 and 250 millimols per liter, respectively). It contains little calcium, magnesium, sulfate or bicarbonate, and the urea concentration is only one-twentieth that of the plasma. (In its isosmotic nature, the rectal gland secretion differs from that of the nasal gland of the marine birds and reptiles, which has nearly twice the osmotic pressure of the plasma.) The maximal rate of secretion in untreated dogfish under experimental conditions is about 1.3 cc. per kg. per hour, a figure approaching or exceeding the simultaneous urine flow.

This rectal 'salt' gland affords a mechanism by which the dogfish can dispose of excess sodium chloride after the ingestion of sea water—the magnesium, calcium, and sulfate, being poorly absorbed from the intestinal tract, are presumably largely evacuated with the feces, and excess potassium may be excreted by the kidneys or gills. The gland appears to serve only as a protective device, like the nasal gland in the marine birds and reptiles; the present evidence does not indicate that the marine elasmobranchs habitually drink sea water, as do the marine teleosts. Added to the urea-retention habitus, it guarantees the animal a continuous supply of water for the formation of an osmotically dilute urine.

It is, however, not enough for the adult of a species to solve any physiological problem unless it is also solved for the embryo, and the urea-retention habitus requires that the elasmobranch protect the egg and embryo against the osmotic loss of water until such time as the embryonic respiratory membranes and kidney develop to the point where they are self-sufficient in respect to the maintenance of water balance. To this end the elasmobranchs first covered the egg with a waterproof ('cleidoic' = closed) case, secreted by a gland low down in the oviduct. This case was usually supplied with coiled tentacles to facilitate its attachment to well-ventilated seaweed. (Empty skate-egg cases are familiar sights along the American shores of the North Atlantic.) A large, urea-rich yolk sac and a sort of artificial lake, equivalent to the amniotic fluid that bathes the embryos of reptiles, birds, and mammals, is thus provided for the young until they develop biochemical independence.

All observers are agreed that oviparous reproduction by means of the cleidoic egg is the primitive mode among the elasmobranchs, though it persists in only a few recent forms; in most families the egg is retained in the oviduct of the mother until the embryo is mature, the case being reduced to a thin diaphanous membrane or replaced by a placenta of sorts, thus giving way to 'ovoviviparous' reproduction. Many curious specializations are found for transferring nourishment to the embryo, ranging from a pseudoplacenta in certain sharks (Carcharinoida) to vascular channels passing from the mother's oviduct to the embryo's gills or intestinal tract through the embryo's spiracles in certain of the rays (Myliobatidae).

But to enclose the egg in a waterproof covering, or secondarily to develop ovoviviparous reproduction, requires that the egg be fertilized within the body of the female, so that for the first time in vertebrate evolution the organism had to resort to internal fertilization. In order to effect internal fertilization, pelvic fins of some sort had to be available to the male for the intromission of sperm into the cloaca of the female: in recent elasmobranchs these pelvic fins take the form of specialized claspers carrying erectile tissue and so designed that they can be inserted into the cloaca during copulation. Pelvic fins and claspers are generally preserved in the fossil record and it should be possible from a study of this record to discover when the uremic habitus, with its dependent mode of internal fertilization, was evolved. The paleontologist, however, has afforded us little information on the point. True claspers are recorded in all the Jurassic sharks, but among older forms they are described only in the Cochliodontidae, which were ancestral to the chimaeroids, and in the fresh-water Pleuracanthodii. Since both of these groups were derived from a common stem (*Cladodus*), and in view of the fact that the fine line between 'pelvic fins' and 'claspers' has not been considered by the paleontologist, and since what are recognized as 'pelvic fins' in the Carboniferous forms may actually have been used as intromittent organs, we infer that internal fertilization may go back to the late Devonian, as suggested in Figure 6. In this view the urea-retention habitus was probably the signal adaptation that enabled the Stegoselachii (which are believed to be ancestral to the Devonian elasmobranchs) to effect the first vertebrate invasion of the sea. By this interpretation, the Carboniferous Pleuracanthodii, which were inhabitants of fresh water and which possessed claspers, must have returned to that habitat after the urea-retention habitus had been acquired.

There is no reason to believe that the elasmobranchs have ever been irrevocably bound to salt water: the list of those known to inhabit fresh water today, either temporarily or permanently, comprises at least 13 families, 19 genera, and 22 species, and includes representative sharks, rays, and skates. Experimental animals transferred to moderately diluted sea water show a marked increase in urine flow, as is to be expected from the inevitable increase in absorption of water through the gills, but species habituated to salt water have not been successfully established in fresh water under aquarium conditions, possibly because the transition must be effected gradually. However, many of the naturally occurring fresh-water forms are identical with marine species and can probably migrate freely back and forth. This is unquestionably true of the fresh-water sawfish *Pristis microdon,* which belongs technically among the sharks. In 1930 the writer was able to study this sawfish in some detail at Teluk Anson, Perak, in the Federated Malay States.

Teluk Anson is about forty miles inland from the mouth of the Perak River where it flows into the Straits of Malacca. Survey of the salinity of bottom and surface water showed that the lighter fresh water tends to float out on the surface of the Straits, while at high tide the salt water shelves inland on the river bottom. At the time of the survey, however, the bottom water showed no significant salt some thirty miles below our Teluk Anson station. *Pristis* occurs in considerable numbers at Teluk Anson and for fifty miles up-river, frequenting the shallower water and growing to a size of sixty pounds, and there is every reason to believe that it is thoroughly habituated to life in fresh water and can reproduce there.

Small sawfish (three to ten pounds) were caught by hand nets in shallow water and transported to the station in spindle-shaped bamboo baskets, which the Malays use to transport live fish from the fishing grounds to the village, towing them behind the sampan, and which they also use to store the fish for periods of a week or more. At least four species of rays and three of sharks at one time or another invade the Perak River to or above Teluk Anson, but the only ones other than *Pristis* taken at this time were the ray *Dasyatis uarnak* and the shark *Carcharhinus melanopterus,* neither in sufficient numbers to permit experimental studies.

In fresh water, the urine flow in *Pristis* is large, aver-

aging 250 cc. per kg. per day, as compared with 12
cc. or less in marine forms. The concentration of urea in
the blood of *Pristis* and in the other fresh-water elasmobranchs studied averages about 70 per cent below the
figure typically observed in marine forms; this reduction
appears to be related to the increased urine flow resulting from life in fresh water rather than to increased permeability of the gills. From these and other data one
can construct the cycle of events as *Pristis* moves up-
river. When it enters the brackish water, it is charged
with urea to the full extent, but it is in only a slightly
superior osmotic position in relation to its environment
—in other words it has only limited quantities of free
water available to it, and the urine flow is at a low level.
As it encounters water of decreasing salinity, its osmotic
position is improved and more water is absorbed through
the gills, with a consequent increase in urine flow; but
as the urine flow increases, the blood urea concentration
is decreased by urinary loss, until a new steady state is
reached. If it turns back to sea, the cycle is reversed and
decreasing availability of water decreases the urine flow
and increases the urea content of the blood until the
animal comes back into water balance. It is a complete
misreading of the record to say that the reduction of the
blood urea content in the fresh-water elasmobranchs is
a 'reversal of evolution': evolution has nothing to do with
Pristis or any elasmobranch swimming up and down a
tropical river—the problem is simply one of an automatic
physiological adjustment.

After diversification in the Carboniferous, the elasmobranchs showed great reduction in number and variety
at the end of the Permian, a crisis that saw the extinction
of many ancient forms. Expansion occurred again in the
Triassic and Jurassic, and by the close of the latter
period nearly all the modern lines of sharks, rays, and
skates were represented in the seas. They survive today
as the oldest order of fishes above the cyclostomes, and

as representatives of the first successful effort by the jaw-bearing vertebrates to penetrate the sea or to solve the problem of water balance away from the fresh water in which they had been evolved.

CHAPTER VI

THE LUNGFISH

When the sharks sought sanctuary from the climatic vicissitudes of the Devonian continents by turning to the sea, they condemned themselves and their progeny to perpetual existence as fishes. Other fresh-water fishes, who took what was perhaps a more dangerous and certainly a more difficult course, are of more immediate interest to those who can read the epic of life in books.

Along with the elasmobranchs, the Devonian bony fishes had inherited many features of the placoderms and especially of the spiny sharks, and at least some of them had deviated from the elasmobranchs in one important respect—they had begun swallowing air as an accessory mode of respiration. Appearing alongside the early Devonian ostracoderms and placoderms, and before the sharks, these air-breathing fishes had by the Middle Devonian given rise to two groups, one of which (the Actinopterygii) was to lead to the modern fishes; the other (the Crossopterygii) to the ancestors of the air-breathing, four-footed animals.

It is presumed that the ancestors of the air-breathing fishes at first simply swallowed air and either passed it through the intestinal tract or regurgitated it, but shortly they developed an air bladder or 'lung' which opened as a blind sac off the ventral side of the esophagus. In the higher fishes this air bladder was ultimately to be converted into the swim bladder, and its use as a respiratory organ abandoned, but among certain of the Devonian fishes the air bladder proved to be a lifesaving device. It persisted as a respiratory organ in two lines: the lungfishes or Dipnoi (*di* = double; *pnein* = to breathe), and the Crossopterygii (tassel-finned fishes), which between them became the most common freshwater fishes in the Middle and Upper Devonian. No crossopterygian survives today, but some insight can be gained into the life of the Devonian air-breathing fishes from studies of the surviving lungfishes, which are closely related to them.

Once numerous and widely distributed in the fresh waters of the Paleozoic continents, the lungfishes are today reduced to three genera: *Neoceratodus*, with one species in Australia, *Lepidosiren*, with one species in South America, and *Protopterus*, with three species in equatorial Africa (Figure 7). The last two are fairly abundant, but *Neoceratodus* is so nearly extinct that the government has placed it under its protection. *Lepidosiren* and *Protopterus* undertake estivation (summer sleep) in the dry season, *Neoceratodus* does not. In *Lepidosiren* and *Protopterus* the gills are vestigial and the animals are so dependent on aerial respiration that they are quickly asphyxiated if restrained under water, while *Neoceratodus* retains functionally adequate gills and can remain for indefinite periods under water, and is reported to die if removed to air.

During some years from 1928 onward the writer was able to carry out extensive studies on the African lungfish, *Protopterus aethiopicus* (called *Kamongo* in Swahili), both during estivation and in the active state. This family inhabits the River Gambia, the Congo basin and the rivers and lakes of equatorial East Africa, and is among the common fishes eaten by the natives around Lake Victoria. The natives catch it in nets in both deep and shallow water, and also seek it out when it is estivating by probing every suspicious-looking hole in the mud with a pointed stick, and, if the end of the stick smells of fish, disinter the victim. It grows to a large size, a specimen in the Nairobi Museum being seven feet long and weighing ninety pounds, though the majority of specimens range up to only two to three feet. It is much more elongated than most of its Devonian ancestors, having somewhat the appearance of a large eel, while its paired fins have lost all the ancient crossopterygian characters and are reduced to long filaments which it uses only for balancing. In water, *Protopterus* lives very much as any other fish, except that it rises to the surface at ten- to fifteen-minute intervals to empty its lungs and gulp fresh air, which is passed into the lung by swallowing; the excess escapes from the mouth after the fish sinks below the surface. It lives chiefly on snails, the shells of which it can easily crush with its flat teeth and powerful jaws.

The equatorial region of Africa is subject to heavy spring rains which alternate with protracted periods of aridity. The severity of the annual rain is itself subject to cycles that correlate with climatological changes elsewhere in the tropics and are probably related to the well-known eleven-year sunspot cycle. Consequently during periods of light rainfall large areas along the shores of the lakes and rivers may remain exposed as mud flats for several years, receiving only a superficial wetting in the rainy season. We had hoped to study the estivating fish in its native habitat but, ignorant of the eleven-year cycle in the rains, we found the country flooded and offering no hope of dry mud for several years. This was, in a way, fortunate, because it forced us to bring active fish back to New York where they could be studied under controlled laboratory conditions for a period of several years. Small specimens, up to a foot or more in length, were readily collected for export from among the papyrus roots at Kisumu on the eastern shore of Lake Victoria. The fact that petrol and oil are imported into Central Africa in five-gallon and one-gallon tins, respectively, neatly packed in small wooden boxes, helped us to solve the problem of transportation. By cutting out the tops of the tins and adding hinges and handles to the boxes, we contrived shipping cases that would pass official eyes in railroads, steamers, taxis, and customs barriers, including those of Egypt and France. We started for New York with thirty-two one-gallon tins of mud, on which a small amount of water was standing, and eight five-gallon tins containing about six inches of water. Two or three fish were placed in each of the tins of mud, and six to ten smaller fish in each of the tins of water—a total of about one hundred and fifty fish.

The weather was unusually cold the first night as the train passed over Mau Summit (8322 feet) on the way to Nairobi, and about forty of the fish, some of which had been placed in the line of draft from the ventilators, were killed by undercooling. This unfortunate start spoiled an otherwise near-perfect record: with the exception of one fish, which was found dead in the mud a few days out of Mombasa—and which, from the extent of decomposition, appeared to have been dead a long time—there were no other casualties until after the fish were in New York and divided between the New York Aquarium and the laboratory. The period of transport was about six weeks, the total period of storage in tin containers some twelve weeks, and the distance via Nairobi, Mombasa, Port Said, Marseilles, Paris, and Le Havre to New York, over eight thousand miles.

These facts all certify the remarkable hardihood of the lungfish. They had suffered repeated splashing and jarring; those in water endured many changes of water including slightly chlorinated water at Nairobi, ship's distilled water at sea, and miscellaneous waters taken at various ports; they endured prolonged and excessively high temperatures in the Red Sea, and one box of mud fell several feet and was turned upside down in a French baggage car—and they were repeatedly poked at by customs officials and taxi drivers. They had remained for weeks in tins, which rusted in spite of a coating of paraffin, and which in many instances showed traces of floating oil. During this time they were unfed, since we were convinced that they would travel better if putre-factive contamination of the mud and water was kept to a minimum. In short, they suffered an ordeal which few other animals, and certainly few fishes, could have en-dured, with only a single casualty attributable to hard-ship other than undercooling.

Part of the fish were placed in the New York Aquar-ium, some in 'balanced' aquaria and some in running fresh water. Those in the balanced aquaria died within several weeks, presumably from infection. Those kept in running water gradually began to eat beef-heart, aban-doned their carnivorous attacks on their comrades, and for three months grew with surprising rapidity until a broken aquarium window required their temporary re-moval to another tank containing a single specimen of *Lepidosiren*. Within a few days the *Lepidosiren* and all the lungfish died of an infection that ran an unusually rapid course (this was before the days of antibiotics). Two fish, however, which passed to the University of Chicago and thence to Dr. Caryl P. Haskins, survive today—twenty-eight years later—in the Aquarium of the New York Zoological Society, and are now almost too large to be housed in the available facilities. They would be larger if the tanks were larger—because, as is known to aquarists, a fish stops growing when its home ceases to fit it comfortably.

The fish that were kept in the laboratory each had an aquarium to itself and infection presented less difficulty. They soon ate well and began to grow, and when they were well established they were put into estivation by simply dropping them into 12 × 12 × 15-inch glass bat-tery jars filled with wet black mud of a fine-grained quality from upstate New York, chosen because it looked like the mud around Lake Victoria. The mud was allowed to dry in a cabinet with a minimal temperature control set at 68° F., and within a month or so it was superficially dry and hard. The choice of mud was a little precious and in many instances the fish failed to survive simply because we did not know how to make bricks: we overlooked the roots, decaying stems and general debris of a natural swamp. As the months passed the mud in the jars began to crack, and if the crack penetrated to the cocoon it tore this protective envelope and the fish, exposed to air, dried out and died. But some two dozen estivating fish survived a year or more to become the casualties of physiological investigation. Since accurate metabolic data could not be obtained while the animal was in the mud, they were removed in the estivating state and imbedded in plaster of Paris with only the nose exposed. Several survived a total period of two years in mud or plaster, and one survived three years of estivation and an additional year of starvation after it was returned to water.

It is clear that the fish trapped in the swamps by the recession of the water in the dry season passes into a state of estivation by a series of responses remarkable for their automatic sequence. As the surface water grows shallow, and while the mud is still soft enough for easy burrowing, the fish squirms into the ooze nose-first and turns upward with the snout just below the surface of the water. With further descent of the water level, by its weight and squirming motions it follows the subsoil water down to a depth of a foot or eighteen inches, and its excursions to the water's surface shape a bulbous cavity that opens to the air through a small blowhole. When the water finally drains away, the fish can at last breathe without moving, and it curls up with its tail across the top of its head, covering the eyes. (Figure 7.) Its body is coated with a slimy mucus secreted by the skin, and as this mucus dries it hardens into a brown, parchmentlike, waterproof cocoon that envelops the body closely, extending into all exposed crevices. The only opening is a short funnel where the cocoon extends between the lips and teeth, and through which the fish breathes.

The lungfish is in no sense a land-living animal, and if exposed naked to the air it will die in twenty-four hours, its skin shriveled like that of a dry frog. But within its cocoon of slime, and excluding minute evaporation through the cocoon, the only route of water loss in the estivating animal is by way of the lungs, and so long as the cocoon remains intact the fish is effectively protected against desiccation. From the time when it nosedives into the mud there is no possibility, of course, of obtaining food, and henceforth it must live on its own tissues. Being a cold-blooded animal, it does not need to maintain a high level of metabolism in order to sustain the body temperature; and consequently, beginning with the first day of fasting, whether active or not, its metabolic rate begins to decrease, dropping by 50 per cent in a week or so, and decreasing slowly thereafter until within three months it reaches the low level of 10 to 15 per cent of that in the fed, active state. This reduction of metabolism during fasting, which is probably characteristic of all cold-blooded animals, is wholly unrelated to estivation as such, to the retention of metabolic waste products, or to the cessation of nervous activity; it appears to result purely and simply from the absence of food and the progressive depletion of the body stores of fat and protein and is perfectly duplicated in fasting but active fish. It is a highly important factor because it serves to increase, four to six times, the period the imprisoned animal can live before it must inevitably starve to death unless it is liberated.

As the metabolic rate and therefore the oxygen requirement decrease with fasting, breathing becomes slower and slower: whereas an active fish breathes at the surface of the water at least every fifteen minutes, the intervals between breaths during estivation extend to one or several hours. The heart slows to perhaps only three beats per minute and, except for the infrequent pulse and respiration, the animal appears to be in a state of suspended animation.

In captivity the lungfish lies quietly on the bottom of the aquarium or leans lackadaisically against the side, moving only to eat or breathe. For the latter purpose it need only flex its body and swim gently to the surface, where the lung is quickly emptied and filled again with air. But it is nevertheless a conscious animal, aware of its environment and of itself. However, when removed from its mud block after several months of estivation, it is as dormant as a sleeping child. One may debate if one can properly speak of 'sleep' in the lower animals—we do not know what sleep is in the higher animals except that it appears to involve cessation of activity in the higher nervous centers—but to use the word loosely, insects 'sleep,' sometimes so soundly that they can be picked up without disturbing them; fishes apparently drowse off for a nap in a quiet corner or under a stone (the mackerel is an exception in that it must keep swimming in order to aerate the gills, so that if it sleeps it does so while swimming), while reptiles, birds, and mammals all have their periodic escape from conscious life.

Whether the estivating lungfish is asleep or not is, for the moment, a matter of a hairsplitting definition—it certainly is not conscious, as is the active animal when it is searching for snails on the rocks and papyrus stems, or following interestedly a piece of beef-heart dangled in the aquarium by teasing fingers. When the cocoon is peeled away, bit by bit, the fish is as moist as though it had just been removed from water, but the body, which gives off a pungent, never-to-be-forgotten odor, is inert except for the fine twitching of a muscle fiber or a sudden respiratory gasp, and it remains so for days or weeks if kept in a moist chamber. This dormant state is not attributable to lowered basal metabolism, dehydra-

tion, or the accumulation of metabolic waste products—
all can be excluded by controlled observations; perhaps
it is an active inhibition induced by prolonged postural
fixation, a sort of autohypnotic ecstasy; or perhaps it is
just a negative state resulting from the absence of any
nervous activity other than that required for respiration,
and engendered by long insulation from any sensory
stimuli.

When the crisis comes, it is the threat of death that
wakens it. Removed from the cocoon and placed in
water, the inactive blob floats until the air that is presently in its lungs is expelled and then it sinks, rolling
inertly to one side or the other on the bottom of the
aquarium. When next it opens its mouth to breathe, the
influx of water shuts the jaws by a reflex that must be a
unique patent of the Devonian air-breathing fishes. After
it has made several futile attempts to breathe under
water (which, of course, it cannot do), asphyxia begins
its work and convulsive movements appear that twitch
and rock the body violently, until, in a culminating par-
oxysm of effort, the animal partially uncurls and strug-
gles with grotesque movements to the surface. We suppose that in nature, when the flood waters cover the
nest, a similar asphyxial convulsion wakens the fish so
that it pushes to the top of its burrow and, breaking the
soft edges of the blowhole, swims free as soon as the
water attains any depth. After its first victorious ascent,
aerial respiration is resumed (we never saw a lungfish
drown if it was not mechanically imprisoned below the
surface) but for several days the stiffened and twisted
body may look very much like an animated horseshoe
trying to stick one end of itself out of the water, and for
days the swimming movements may be erratic and
unco-ordinated, as though the animal had forgotten
some of the most elementary motor acts. If the lungfish
is capable of learning by experience, it seems reasonable
to suppose that this knowledge is all erased during its
long sleep; it can be suspected (though our laboratory

observations afford no proof) that the creature has to
learn how to eat again. Time, as it were, has passed it
by, and one can seriously ask if it has grown older with
the passing years, a speculation that assumes less of the
element of frivolity when we are reminded how little we
know—apart from the onslaughts of the 'degenerative
diseases'—about the process of 'aging' in any species,
including man.

Once free, always free—unless it is again caught in
shallow water at the onset of the dry season. This is a
matter more or less of chance, since the lungfish is
neither a geographer nor a meteorologist and can
scarcely be credited with prescience in such matters.
When it burrows into the mud it does not know how long
it will be imprisoned there: this is a question of winds
and rains and floods, of eleven-year cycles of sunspots.
Neither does it prepare for incarceration with any foresight, since it may be fat or lean when it goes into estivation. Indeed, the lungfish is poorly prepared for prolonged fasting under any circumstances, because it stores
and utilizes relatively little fat, which is the big reserve
of energy in the warm-blooded animals. In no animal
does stored carbohydrate supply much energy during
starvation, and in the lungfish the reserve of carbohydrate is gone in a few days and from then on 50 per
cent of its energy is derived from fat, the other 50 per
cent from tissue proteins, chiefly in the skeletal muscles.
When the fat is gone nothing but muscle protein remains to be burned until even this is excessively reduced
and the animal expires in a final, rapid conflagration
in which the tissues are literally disintegrated. The meager use of fat during fasting appears to be a general
rule among the cold-blooded animals, the storage and
utilization of large quantities of this energy-rich fuel being a concomitant, apparently, of the evolution of the
warm-blooded state in the birds and mammals. A fasting man, for example, derives 85 per cent of his energy
from fat, and only 15 per cent from body protein; conse-

quently per gram of body weight lost in fasting he obtains 4 o calories to the lungfish's 1.4, thus making him a better candidate for fasting, per unit of body weight lost, by a ratio of 2.5 to 1.0. But being warm-blooded, man spends his store of energy in thirty to sixty days, while the lungfish, because it is cold-blooded and because of the remarkable reduction in its metabolic rate in the fasting state, can spread its reserves over several years.

In the dry mud the lungfish has no water available to it except the small quantity of so-called metabolic water formed by the oxidation of carbohydrate, fat and protein, and an additional small quantity liberated from the tissues as they are degraded. Nothing is known about renal function during estivation except that no urine is formed (and it is rather difficult to study any physiological function when it is zero). The cloaca, into which the ureters open, is sealed tight by the cocoon, and there is no evidence of any urine excretion after the cocoon is formed. This would be impossible in any case: we may be certain that the blood flow to the kidneys, as in all other parts of the body, is reduced to minimal maintenance levels; the blood pressure must decrease as the heart slows and the process of glomerular filtration must be wholly suspended. A mere trickle of blood through the renal-portal system and the glomeruli suffices to keep the renal tubules alive.

With excretion wholly arrested, all the nonvolatile products of metabolism must accumulate in the body. The one produced in largest amount is the urea, which is formed by the metabolism of body protein. It has been emphasized that urea is a relatively nontoxic substance, and where it accumulates slowly in the body, as in the estivating lungfish, it probably has no effect whatever on internal salt and water distribution or the function of any organ. At the end of the first year of estivation, the urea content may reach 2 per cent of the total body weight; the highest figure observed in an estivating fish was 3.1 per cent at the end of 1105 days, but if post-estivation metabolism is included, the maximal figure attainable appears to be closer to 4 per cent.

In the combustion of protein the sulfur and phosphorus contained in the protein molecule are normally oxidized to inorganic sulfuric and phosphoric acids and excreted in the urine as neutral salts. Inorganic sulfate accumulates in the estivating lungfish, but phosphate is somehow retained in an organic state so that there is none to be excreted after estivation. There is no accumulation of other notable products of metabolism, no creatine, creatinine, ketone bodies, uric acid, or ammonia. The absence of ammonia is of particular interest, since of all nitrogenous end-products this is one of the most poisonous. Active fish excrete 30 to 70 per cent of their nitrogen as ammonia, almost all the rest appearing as urea, and both are excreted almost entirely by gills; the failure of ammonia to accumulate during estivation conforms with other evidence that the ammonia excreted by the active lungfish (and that excreted by other fishes) is not that which is formed in the body by the metabolism of protein (which is rapidly converted to nontoxic urea by the liver) but is ammonia that is formed *de novo* peripherally by the gills or kidneys from some nonprotein precursor.

Certainly the lungfish, in four years of complete anuria, suffers no autointoxication. We estimate that if the animal has a maximal store of fat when it goes into estivation, and if it gets the breaks from nature (which knows how to make bricks properly) it might survive for seven years—which, even with an eleven-year rain cycle, is a very long drought.

When the dormant fish is returned to water the accumulated urea is excreted in a spectacular manner through the gills, but not fast enough to prevent its transient osmotic action, so that water is rapidly absorbed and for a few days the fish gains markedly in weight and may acquire a swollen, waterlogged appearance. That renal function is quickly re-established is shown, however, by the excretion of sulfate, which is not excreted by the gills. Then within ten days or so the animal, although emaciated in the extreme, is back in excretory balance and ready to start life anew.

Surprisingly, *Protopterus* breeds shortly after it emerges from estivation so that it has scant time for recuperation. The breeding fish make a nest of sorts in the swamp grass in shallow water where the eggs, externally fertilized, are deposited on the bottom. The young have a suctorial organ resembling that seen in tadpoles, by means of which they attach themselves to the sides of the nest, and they have long external, cutaneous gills—both of these larval organs disappearing as the mature state, which ushers in aerial respiration, is reached. Until the eggs are hatched, which requires about eight days, and while the larvae remain in the nest, the male stays on guard and lashes the water with his tail to improve aeration. However, in his ravenous hunger, he may eat all the young.

Air breathing was probably practiced by all the important groups of Devonian fishes except the elasmobranchs, but how many of them undertook estivation is unknown. It is only in recent years that lungfish burrows have been recognized in the Paleozoic sediments of Texas, a few containing the remains of a fish that failed to escape its prison in the mud. We may believe, however, that by means of their aerial respiration the lungfishes survived in swamps fouled with decaying plants, and on occasion in wet mud for short periods, and that their cousins the Crossopterygii, by virtue of their primitive lungs, squirmed across the land from one pool to another, to lay the foundations for the evolution of the first four-footed, air-breathing animals.

It is an interesting postscript to the evolution of lungs that the air bladder continues to be used as a lung only in the Dipnoi and a few primitive forms such as *Polypterus* of the Nile and the bowfin and gar pike of North American rivers. Among most of the higher fishes, this once lifesaving device has been converted to a swim bladder, which may or may not be sealed off (if sealed off it is filled with oxygen secreted from the blood), and which serves as a hydrostatic organ to enable the fish to live at a convenient depth (as in the common trout and salmon); or it has disappeared entirely (except for a transient appearance in the embryo) in many fishes— such as the mackerel, which is specialized for speed, the flounder, which is specialized for bottom life, and the lumpsucker, which is specialized for suctorial attachment to rocks.

CHAPTER VII

THE AMPHIBIA

It was very well for the lungfish to lie dormant for months or years in its mud nest, but it was better, and a new way of life, to keep awake by crawling from one water hole to another; and the evolution of the four-footed animals, or tetrapods, which first emerged from aquatic to semi-terrestrial life, is one of the most notable landmarks in vertebrate history. A few invertebrates, such as the scorpions, millepedes, spiders, and perhaps a few wingless insects, had established themselves on land in the Devonian, but these were small and feeble animals; they subsequently evolved in large variety and numbers, but they failed to achieve the great destiny of the ter-restrial vertebrates.

The evolution of the Amphibia began in the Devo-nian, the aridity of which had fostered aerial respira-tion in the continental fishes, but it might have come to nought had it not been for the more favorable circum-stances of the Carboniferous. This geologic period de-rives its name from the widespread deposits of coal that were then formed, and was so named in England where, in the coal-bearing seams (or 'Coal Measures,' as the English called them), the first half of the period is meagerly preserved. As represented in the United States, however, the 'Carboniferous' is divisible into two dis-tinct and quite dissimilar periods, the Mississippian and the Pennsylvanian.

At the close of the Devonian the earth suffered its third major upheaval in vertebrate history—the Acadian disturbance, named from Acadia of Maritime Canada where the Devonian and underlying sediments remain exposed today as greatly uplifted and tortuously wrinkled strata. The Acadian disturbance culminated in a second generation of Appalachian Mountains running along the line of the older (Ordovician) Taconic range; it is estimated that in this one area alone the Acadian uplift exceeded the volume of the present Sierra Nevada, a range 75 to 100 miles wide and 400 miles in length, and rising to nearly three miles above sea level at its crest. Volcanic activity was marked in southern Quebec, the Gaspé Peninsula, New Brunswick, and Maine, wherever the crust was broken by deep faults, while continued uplift and erosion exposed the basal continental granite in Maine at Mount Katahdin, and in the White Mountains—the latter once rising at least 12,000 and possibly 17,000 feet in height. Early in the interval following the Acadian uplift an inland sea spread over what is now the Mississippi Valley and there laid down the rich fossil record which gives to the early American Carboniferous its name.

The Ouachita disturbance, which closed the Mississippian, is considered by some historical geologists to be the forerunner of, and essentially integral with, the great Appalachian revolution that was ultimately to close the Paleozoic era and bring about profound transformations in all forms of life. Between Mississippian and Pennsylvanian time the eastern and southern parts of North America had again been greatly elevated, and mountains had begun to rise in Colorado and South Dakota, to form the Colorado Mountains (the Rockies are of more recent origin); so that near Leadville, Colorado, for example, marine limestone which was laid down in the Mississippian now lies two miles above sea level. But even as the eastern and western edges of the continent rose and buckled into mountain chains, the central basin sank ir-

regularly below the sea to be overlaid with rich alluvial
soil to a depth of several thousand feet, forming three
great deltas, one in east-central Pennsylvania, another
in West Virginia, and a third in northern Alabama. As
these alluvial deposits grew in depth, the shallow sea
became broken into bayous, and ultimately the salt
water was displaced by fresh-water streams and inter-
lacing lakes, to give rise to the vast swamps which were
to be transformed into coal. Because it was in this period
that the great coal deposits were formed all over the
world, and because it is in Pennsylvania that the greatest
coal veins in the Western Hemisphere are found, the
period is designated the Pennsylvanian by American
geologists.

The Mississippian climate of Europe and North
America was for the most part a semiarid one, relieved
only by short if intense seasonal rains. Large areas of
both hemispheres were covered with landlocked salt
lakes in which deep beds of salt were deposited, or by
deserts that spread from the subtropical regions into the
present Temperate Zones. On the whole it was not a
climate conducive to the multiplication and diversification of the Devonian terrestrial forms, though the tetrapods left traces of their existence in the form of
footprints in the Upper Mississippian red beds of Pennsylvania—associated with mud cracks and marks of the
raindrops presaging the short-lived flood that later
buried them—perhaps a record of a tragic search for
water as the mudholes gave way to hard-baked flats in
the summer sun.

The Pennsylvanian, however, saw a remarkable transformation in climate. The annual mean temperature of
Europe and North America is estimated to have risen to
53° F., or 20° higher than it is today, partly because the
spread of oceanic currents carried warm tropical water
so far into the Arctic region that marine corals grew as
far north as the now glacier-capped island of Spits-

bergen. In this moist climate tree ferns, seed ferns, and other plants, very much alike all over the world, developed with unprecedented luxuriance. Characteristic of the Coal Measures everywhere were the giant scouring rushes—of which the diminutive horsetail, *Equisetum,* is a humble descendant—growing to a height of 30 feet and a diameter of 12 inches. The largest plants were the 'scale trees,' the stumps of which attained a diameter of four to six feet and the trunks a height of 100 feet. Almost as large were the Cordaites, the forerunners of modern conifers, and the only trees that to a modern eye would look even superficially like trees; their leaves, however, were blades and not needles, and their seeds resembled bunches of grapes rather than pine cones. (There were, as yet, no flowering plants, most of which were not evolved until the Cenozoic era.) The roots of this lush vegetation were immersed in standing water, as in the present Florida Everglades and in the bayous of the Mississippi River, and as the vegetation died it sank below the surface, where it was protected from decay, generation upon generation piling up until the coal seams reached a depth of thirty-five to forty feet.

The Pennsylvanian was probably the warmest period associated with abundant rainfall in the history of the earth, and in this agreeable climate terrestrial animal life also came into its own. Insects underwent rapid evolution and grew to enormous size: dragonflies had a wingspread of twenty-nine inches, and the cockroaches were so big—three to four inches long—that the age has facetiously been called 'the Age of Cockroaches,' though it would be fairer to call it 'the Age of Insects' were it not that it is already known as 'the Age of Coal.' Spiders, twelve-inch centipedes, snails, and hundreds of species of scorpions luxuriated in a bountiful nature and supplied abundant food for the great Amphibia whose tracks, but not whose bones, are preserved in the preceding Mississippian record.

The parting of the ways that separated the first terrestrial animals, the Devonian Amphibia, from the air-breathing fishes depended not so much on the efficiency of aerial respiration as on the structure of the fin that was to become a foot.

A fish propels itself primarily by means of movements of its tail and body, the fins serving chiefly to maintain an even keel, to prevent pitching, and to improve the accuracy and speed of turns. In the Devonian fishes the pectoral and pelvic spines had become connected with the underlying muscles and articulated with the pectoral and pelvic girdles, to supply one pair each of pectoral and pelvic fins as these appear in the higher fishes. Professor Romer has remarked that there is nothing sacrosanct about four as a limb count, and it is amusing to speculate about the possible results had some of the spiny sharks, with up to seven pairs of ventral appendages, survived in the higher animals. (Again, science fiction writers may take note!)

Among the Devonian bony fishes the fins had lost the spines of the placoderms and acanthodians and developed a basic pattern, which had started to evolve in two directions. In one group the fin developed many fanlike rays anchored to a long base, thus giving rise to the 'ray-finned' fishes (Actinopterygii, *aktinos* = ray; *pteron* = wing; *pterygion* = little wing or fin) as represented by the common fishes of today. This many-rayed fin was excellent for swimming—but it was too feeble to support the weight of the animal out of water, and very few of the Actinopterygii could crawl even short distances on land. In the few ray-finned fishes that essay to scramble out of water today—such as the mudskipper, *Periophthalmus*, and the Indian climbing perch, *Anabas*—the fins have secondarily been remodeled into relatively feeble crawling or climbing appendages.

In the other group, however, the primitive fin developed into a more compact paddle- or tassel-like structure, giving rise to the Crossopterygii (*krossoi* = tassels;

pterygion = fin), which had strengthened the bony sup-
ports and muscles and developed a ball-and-socket joint
that permitted the fin to be moved in a figure-eight motion, until all four fins came to be rotated under the body,
lifting the belly clear of the ground. It was in some form
closely allied to the Devonian crossopterygian, *Eusthenopteron* (see Figure 7) that the vertebrates first
crawled out of the water onto the land. The Crossopterygii were first cousins to the Devonian lungfishes, but
the lungfishes, despite their aerial respiration, must be
excluded from a place in tetrapod evolution because of
their specialized elongate fins.

The *Eusthenopteron* stem also gave rise to the Coelacan-
thini, fairly large fresh-water fishes in which a narrow
fin base supported a wide web, supplying an adequate
fin for swimming but one that was too weak to support
the weight of the body out of water. Like the other
Devonian and Mississippian fishes, the coelacanths probably breathed air, but in the Triassic they migrated into
the sea, and the air bladder, no longer used for respiration, became calcified. They disappeared from the fossil
record in the Cretaceous, some 75,000,000 years ago,
and until recently it was accepted that they had become
extinct sometime in that period. Then, in December of
1938, a trawler working off East London in Cape Prov-
ince, South Africa, at a depth of about forty fathoms,
brought in a living survivor of this ancient race. This
specimen, named *Latimeria chalumniae,* was five feet
long and weighed one hundred and twenty-seven
pounds, and was described as steel-blue in color with
dark blue eyes. The find was reported to Miss Courtenay-
Latimer, Curator of the East London Museum, who
communicated at once with Professor J. L. B. Smith of
Rhodes University College, Grahamstown, but unfortunately her letter was delayed for ten days and meanwhile
the body had putrefied and had been disposed of, and

only the skin and head were saved by the local taxi-dermist.

After an intensive search of fourteen years a second coelacanth (named by Professor Smith, *Malania an-jouanae*) was caught off the coast of the Island of An-jouan in the Comoro group, two hundred miles west of Madagascar, in December, 1952. Subsequently numer-ous living specimens weighing from 70 to 180 pounds have been collected in the off-shore waters near the Comoro Islands by French zoologists, and have come to intensive study under Professor Jacques Millot, Director of the Institute of Scientific Research of Madagascar.

The living coelacanths are crossopterygians, differing from Devonian and later forms only in minor details. One notable feature is that the lobate pectoral and pelvic fins have a wide range of movement; the pelvic fins in par-ticular can be rotated under the body, in a position to support, partially at least, the weight of the fish, though actually they do not appear to be strong enough for this purpose. Present specimens have been taken at depths of 80 to 200 fathoms, but Millot suggests that the fish probably lives at a depth of 400 fathoms. During the course of their long history the coelacanths spread all over the world, living in ponds, lakes, streams, coal-swamps, and shallow epicontinental seas. Though a few invaded the sea in the Devonian, the group as a whole continued in fresh water until the Triassic; how long *Latimeria* and its ancestors have been marine is unknown —if absence from the fresh-water fossil record is evi-dence, ever since the close of the Triassic. Perhaps throughout most of the Cenozoic the coelacanths have struggled along as a dwindling group, living along the continental shelves in the manner of rock fish.

Unfortunately many details of interest to us here are not yet known—whether the fish is ovoviviparous or pos-sesses a cleidoic egg, whether the pelvic fins can act as claspers, and whether urea or some other organic, os-motically active solute is present in the blood in exceptional amounts. The paired kidneys are fused, and the organ is anomalously applied to the ventral, rather than the dorsal, wall of the abdomen. Two ureters are present, and each expands into a voluminous ear-shaped bladder. The kidney possesses pine-cone-shaped glomeruli and contains 'cavities lined with ciliated epithelium, which are probably nephridial but the disposition of which I have not yet been able to work out' (writes Millot).

A single coelacanth fossil from the marine Jurassic fossil beds of Solenhofen, Bavaria, presents a full-grown fish containing within the body cavity the skeletons of two small individuals of the same species. D. M. S. Watson, after considering the possibility that these small fishes may represent food, rejects this interpretation in favor of the view that they are unborn embryos and that this particular coelacanth was ovoviviparous. The evidence, however, falls short of proof. Alternatively, the coelacanths may have adapted themselves to salt water in the manner of the recent marine teleosts—that is, by drinking sea water and excreting the salt out of the gills —but in this case we would expect them to have an aglomerular kidney like that of many of the bony fishes (Teleostei) discussed in Chapter VIII. We must await further reports on *Latimeria* before we can understand how this living fossil invaded the sea.

The record of the Amphibia in the Devonian, the period that saw their early evolution, is very sparse—an incomplete skull from the famous Upper Devonian beds at Scaumenac Bay in Canada and a few skulls from freshwater deposits in Greenland which may be either late Devonian or early Mississippian in origin. These were fair-sized animals with skulls half a foot or so in length, and probably represent an advanced stage in amphibian evolution. But in the succeeding Coal Measures of North America seven amphibian orders, comprising 19 families, 46 genera, and 88 species, have been identified. Most of these were only a few inches long, but a few such as *Eogyrinus* (see Figure 8) reached a length of fifteen feet, and one, *Onychopus gigas*, which has left us only its footprints, had a stride of thirty inches and is estimated to have weighed six hundred pounds. Most of the Carboniferous Amphibia were, however, small salamanderlike creatures with a well-developed tail (the tailless frog is highly specialized for jumping), and they moved by twisting the body in fishlike motions, using the legs as pivots and letting the tail drag on the ground. *Microbatrachus* and *Miobatrachus*, only a few inches long, were apparently ancestral to surviving forms. (Some centuries ago a large fossil salamander closely related to the living hellbender, *Cryptobranchus*, was recovered from Miocene deposits near Lake Constance and was identified as the remains of a poor human sinner drowned in the Flood, and called *Homo diluvii testis*. This precise relationship is far off the mark, but the ancestors of this salamander do stand in the line of man's descent.)

By the Permian the Amphibia had become better adapted to life on land: *Eryops*, five feet in length, had short but powerful legs, and *Cacops*, only sixteen inches long, was armored with dermal plates along the back, presumably as a defense against the carnivorous reptiles which had then become abundant. *Seymouria*, from the Lower Permian red beds of Texas, reached a length of twenty inches and stands close to the ancestral line of the reptiles and mammals, and it may have laid a shelled egg on dry land.

The Amphibia were, however, neither fish nor good red flesh. With the possible exception of *Seymouria*, they continued to lay their fishlike, naked eggs in shallow water or in moist places, and the young hatched in the form of fishlike larvae with fishlike gills, to spend the first part of their life in water, only later metamorphosing into the adult form under the endocrine control of the pituitary and thyroid glands. As larvae, they probably fed on small aquatic plants and animals; as adults,

on fish, the larvae of other amphibians, or the giant cockroaches, dragonflies and other insects that inhabited all
the swamps. They reached their greatest expansion in
the Pennsylvanian and began to dwindle in numbers
from the Permian onward, perhaps in part because of
competition with the reptiles that came to their ascendancy in the late Mesozoic, in part because of climatic
conditions unfavorable to their mode of life. After the
Triassic only two main lines survived: one that led to
the frogs or toads that have secondarily lost the tail, and
hence are called the Anura (*an* = without; *oura* = tail),
and a second line that led to the salamanders or Urodeles
(*oura* = tail; *delos* = evident). A third and rare group
survives in the Apoda (*a* = without; *poda* = feet), small
wormlike creatures that live in the moist earth.

To convert a mud-crawling fish into an amphibious animal required anatomical transformations throughout the
body—and something between fifty and one hundred million years. In a fish, the blood goes directly from the
heart to the gills by the ventral aorta, and from the gills
it is distributed by the dorsal aorta to all parts of the
body. When the lung took precedence over the gills as
a means of respiration, the blood had to be diverted from
the gills to the lung by reconstruction of the aortic circulation and the partial division of the heart into two
chambers, a division that was not to be completed until
the evolution of the birds and mammals. The pectoral
and pelvic arches were strengthened to support the legs,
and the vertebral column was transformed into a strong
but flexible arch capable of lifting the abdomen more
or less free of the ground. The body for the most part
lost the scales of the ancestral fishes except on the
belly (which needed protection against rubbing on the
ground) and became covered with moist skin that came
to play an important part in salt and water exchange
and in respiration, and this skin had to be periodically
replaced by molting, in order to keep it viable. The eye

had to be remodeled for vision in air, with the consequence that the brain became increasingly concerned
with the sense of sight. For the first time the animal
began to breathe through the nose, thus bringing the
'nose-brain' into a role of greater importance. (It was
from this 'nose-brain' that the cerebral hemispheres of
the mammals were to be evolved at a later time.) And
so it was with almost every organ—some greater or lesser
transformation contributed to this, one of the most revolutionary events in vertebrate history.

It is something of a paradox, in view of the many other
structural changes in this class, that the evolution of the
Amphibia saw no marked change in the structure of the
kidney: in all recent forms the glomeruli remain large
and well vascularized and the tubule consists, as in the
fresh-water fishes, of the typical proximal and distal segments. There can be no doubt that the Carboniferous
Amphibia, like recent forms, were incapable of elaborat-
ing a urine osmotically more concentrated than the
blood, and it was this circumstance, as much as anything, that limited their habitat. Like living salamanders
and frogs, they had to spend most of their lives in the
water or at best, like the toads, not too far away on
moist ground or in a moist nest of moss or leaves. Nevertheless, they added two patents which served the ends
of water balance. first, a mechanism for reducing the
rate of glomerular filtration (and hence the excretion of
water) when they were out of water; and second, a
mechanism for varying the permeability of the skin to
water. Both of these mechanisms are governed by the
pituitary gland.
Occasion has not hitherto arisen to speak of any of
the glands of internal secretion (the endocrine glands)
which are so important in the physiological activities of
the higher animals, not because such glands are lacking
in the fishes but because so little is known about them.
However, one of these glands, the pituitary, which is

importantly concerned in the regulation of the composition of the internal environment in all the higher animals, appears to have first assumed this role in the
Amphibia.

The pituitary is not only the most important gland of
internal secretion in the body, but it is also the most
complex, because it is composed of several different
types of tissue, each having a different physiological
function. All parts of the pituitary are of ancient origin,
but there is little information on their specific function
below the mammals, and almost no information on their
evolution. The most we can say is that a 'pituitary gland'
is present in all vertebrates. In the cyclostomes, the lowest living vertebrates, the pituitary is connected by nerve
fibers with the 'pineal eye' in the top of the head. This
pineal eye remains something of a mystery: it was present in certain ostracoderms (cephalaspids and pteraspids) and in a few Mesozoic Amphibia, while it was
particularly prominent in some of the Mesozoic reptiles;
it persists, in a degenerate form, in the living archaic
lizard, *Sphenodon*, though here it has lost its visual function as it had probably lost that function in the Mesozoic
animals. In the elasmobranchs and teleosts what is called
the 'pineal organ' is no longer an eye and presents no
trace of photoreceptor structure, and in the mammals
it appears as the 'pineal gland,' the function of which
(if any) is unknown.

The pineal eye had doubtless functioned as an 'eye'
in the ostracoderms, serving to guide these bottom-living
animals through the lights and shadows of relatively
deep water. Of greater interest is the fact that it was
connected by nerve fibers with the pituitary gland, which
secretes, among many other hormones, one that causes
expansion of the amoeba-like cells (chromatophores) in
the skin that are charged with pigment (most frequently
the black pigment, melanin); it is largely by contraction
or expansion of these chromatophores that the lower animals change their color, and it is possible that the pineal

eye was primitively a light-sensitive device enabling the
ostracoderms to assume protective coloration. Long after
the pineal eye disappeared the pituitary continued to
secrete a chromatophore-expanding hormone—in the
elasmobranchs, teleosts, Amphibia, and reptiles, despite
the fact that in many of these higher forms a finer and
more rapid nervous control of the chromatophores is superimposed on or completely replaces this older endocrine regulation. Paradoxically, in the birds and
mammals that have absolutely no chromatophores, the
pituitary continues to manufacture large quantities of
this chromatophore-expanding hormone, but it is conceivable that this hormone now serves some other but
still unknown function in these classes.

The control of protective coloration was, however, certainly not the only function of the primitive pituitary,
because even in the cyclostomes the gland is differentiated into several distinctly different types of cells and
must have subserved several different functions. But here
we encounter a big hiatus in knowledge, and must rely
on inferences drawn from the function of the gland in
the higher animals. In addition to secreting several hormones important in controlling the growth and metabolic
activity of other glands and tissues, the mammalian
pituitary secretes two or more hormones important in
salt and water balance. The best known of these is the
'antidiuretic' hormone (familiarly known to renal physiologists as ADH), which is secreted by that part of the
gland called the neural lobe or neurohypophysis. This
hormone derives its name from the fact that when administered to man and other mammals in very minute
doses it prevents water diuresis by promoting the tubular
reabsorption of water from the glomerular filtrate; in
minimal, physiologically effective doses it has no effect
whatever on the filtration rate.

In the Amphibia pituitary extracts not only enhance
the tubular reabsorption of water but in larger doses decrease the filtration rate by constricting the glomerular

arterioles, thus reducing the filtration pressure. When excess water is available (as when a frog is partially immersed in water), the filtration rate is increased and water reabsorption is decreased, thus increasing the excretion of water; when the animal is out of water and is suffering some dehydration, the filtration rate is reduced and water reabsorption is increased, thus conserving water. In the latter circumstance, tubular function is sustained, despite the reduction in glomerular circulation, by the blood coming to the kidney by way of the renal-portal circulation.

In the mammals (the only forms that have been carefully studied in this respect), the secretion of ADH into the blood is controlled through osmotically sensitive cells located in the midbrain (hypothalamus) and connected by nerve fibers with the neural lobe of the pituitary; the effective stimulus that excites these 'osmoreceptors' is the osmotic pressure of the plasma (which is, of course, determined primarily by the sodium chloride concentration), and so sensitive are they that an increase in concentration so small that it is scarcely detectable by chemical means suffices to induce maximal ADH secretion. It is assumed that increased secretion of the effective pituitary factor in the Amphibia is similarly elicited by increased concentration of the blood when the animal, sitting too long on a lily pad, has lost water by evaporation from the lungs and skin and in the excretion of urine; and that decreased secretion of the pituitary factor is conversely related to excess hydration, as when the animal has been too long in the water, with only its nose exposed to air. But these are areas that remain for future investigation.

When the Amphibia abandoned the scales of their piscine ancestors in favor of a naked skin they encountered a double hazard—excessive hydration of the body when they were immersed in water, and excessive dehydration when they left it. Against these hazards they evolved an additional defense in the form of a mechanism, as yet poorly understood, by which water absorption through the skin can be functionally increased or reduced. This mechanism is under the control of another pituitary hormone, the 'water-balance' hormone, which acts to increase the rate of water absorption so that a dehydrated frog or toad can rehydrate itself quickly when it returns to water, perhaps even to a moist nest, and yet does not lose water excessively when in the air. The absorption of water through the skin is apparently not an active process, but one depending solely on the passive diffusion of water, and therefore it is inferred that the pituitary 'water-balance' principle works by increasing the permeability of the skin to water. It is now generally accepted that changes in permeability are effected by changes in the size of 'pores' through which water can freely diffuse, but the nature of these 'pores' is wholly unknown.

The Amphibia also absorb sodium chloride through the skin, even when this salt is present in faint traces in the external medium; but unlike the absorption of water, salt absorption is an active process and one that probably operates specifically on the sodium ion, to the exclusion of potassium, calcium, and ammonia. There is as yet no evidence that it is under pituitary control. Thus the skin absorbs salt (actively) and water (passively), so that to obtain either salt or water the Amphibia do not need to drink water, which apparently they rarely do.

As in all naked-skin animals, the outermost layer of the epithelium of the skin tends to become hardened or cornified and the exposed cells lose their viability. Where the skin plays an active role in salt and water balance, as in the Amphibia, this process of cornification becomes a handicap—and consequently the animal molts at intervals, casting off the dead outer skin in flakes or large pieces. During molting, both the permeability

of the skin to water and the active absorption of salt
are markedly increased by the removal of this inactive
epithelium. Contrariwise, salt and water absorption must
be considerably decreased in the arid-living toads, where
local thickening of the epidermis into warts foreshadows
the evolution of reptilian scales and plates, and the withdrawal of the skin from active participation in salt and
water balance.

No amphibian can live long in dry air: a frog or newt
exposed at room temperature and humidity will lose one-
quarter of its body weight in ten hours, a degree of desiccation that is lethal for most species. Legends of frogs
or newts recovered alive from the walls of buildings
where they have allegedly been buried for months or
years without water are the product of uncritical observation. No amphibian capable of estivation comparable to that of the lungfishes is known; the nearest approach is among the desert-living frogs and toads that
hibernate during droughts in burrows a foot or so in the
earth. But even here the subterranean air is quite humid
and before hibernation the animal stores considerable
water in the urinary bladder, subcutaneous tissues, and
peritoneal cavity. Some of these hibernating frogs—
for instance *Notaden bennetti, Heleioporus pictus, Cyclorana* (= *Chiroleptes alboguttatus*) of Australia—have
peritoneal funnels (coelomostomes) which drain into
venous sinuses in the kidney, and during hibernation water stored in the peritoneal cavity is thus made available
to the circulation. It is certain that water may also be
reabsorbed from the urinary bladder for this purpose, if
the urine is relatively dilute. So rapidly can the dehydrated Australian frog absorb water through the skin,
a correspondent from Australia once reported to P. A.
Buxton, that 'If you put a lean, dry, herring-gutted
Chiroleptes into a beaker with two inches of water, in
two minutes your frog resembles a somewhat knobby
tennis ball.' It is said that the water-loaded animals are

used by the Australian aborigines as a source of drinking water. Though some of the glomeruli in *Cyclorana* are small and poorly vascularized, there is no evidence of aglomerular tubules, and no aglomerular amphibian is known.

There is a single record of frogs and their tadpoles living in salt water in the Philippines, but it has never been confirmed and it is generally accepted that the Amphibia are incapable of tolerating any degree of salinity above that of their blood. For all of them, past and present, sea water can be considered to be quick poison.

The Amphibia have never conquered the problem of living entirely away from water. It has been said that they are adapted not so much to live on land as to remain in water by getting from one water hole to another. It is not surprising that among many ancient forms (as among some surviving salamanders) the struggle was given up and that the animal has reverted to permanent aquatic life, even to the retention of larval gills, as in the mud puppy and some other urodeles. Nevertheless, the complementary relations between the changes in glomerular filtration and water absorption through the skin, both controlled by the pituitary, serve to keep them in water balance so long as they do not move too far away from the water or moist areas, and, viewed historically, the Carboniferous Amphibia represent the second step—the crossopterygian lung being the first—toward a truly terrestrial life.

CHAPTER VIII

THE BONY FISHES

Though this chapter is broadly entitled 'The Bony Fishes,' it is not our intention to add to the evolutionary history of the fresh-water forms, but rather to consider the fishes that live in the sea. In a sense, the following discussion is a detour both in the progress of our story and in the actual evolution of the vertebrates: we are in the position of the motorist who, when he found himself at the end of a road in the Tennessee mountains, inquired at a hillside cabin how he could get to Nashville; and was told, after considerable cogitation, "Mister, if I wanted to get to Nashville, I wouldn't start from here" The point is, that though it was no mean task for the bony fishes to invade the sea in the first place, having done so they found themselves in a blind alley which did not admit of any great evolutionary advance.

Nevertheless, the marine fishes are of interest in themselves, for several reasons. They afford a substantial fraction of the world's food supply, a no mean consideration. They have also afforded us one of the most interesting chapters in the comparative physiology and anatomy of the kidney. It was the study of the marine fishes that led to the discovery of the aglomerular kidney, and indirectly to a better knowledge of how the kidney works in antecedent glomerular forms; and these studies, in turn, supplied a considerable part of the knowledge through which there were developed some of the important methods for evaluating renal function in man. . The marine fishes also supplied the key ideas which made possible the reinterpretation of the evolutionary history of the kidney. Hence a brief excursus into the blind alley of the sea can scarcely be avoided.

The higher or bony fishes, which populate the seas, oceans, rivers, and lakes of the world today, are known as the Teleostei (*teleos* = complete, perfect; *osteon* = bone) because of the complete calcification of the vertebrae. These bony fishes are represented by some thirty thousand living species and subspecies classified under several thousand genera and six hundred-odd families. No accurate count by families is available, but the vast majority are marine.

Just when the marine teleosts, as represented by surviving forms, became established in that habitat remains a matter of conjecture, though it was certainly long after the elasmobranchs entered the sea. If the predecessors of the recent marine forms invaded salt water in the late Carboniferous, an interval of nearly a hundred million years separated their migration from that of the elasmobranchs; if the marine teleosts have been inhabitants of that medium only during the Cenozoic, this interval is nearer three hundred million years. In any case, the marine elasmobranchs and the marine teleosts are very different from each other in many ways, and particularly in respect to the mechanism by which they maintain water-balance.

Where the elasmobranchs used the urea-retention device to extract water from sea water, the teleosts, evolved from fresh-water fishes, independently and at a much later time, lacked this device. Nevertheless, they had to solve the same problem as did the elasmobranchs: they had to separate water from sea water—which is two and a half times as concentrated as their blood—in order to make a urine no more concentrated than the blood. They solved this problem by drinking sea water and absorbing both the salt and water in the gastrointestinal tract; and then they excreted the bulk of the salt through the gills, leaving free water in the body available for urine formation.

The teleost can live in salt water not because of any superiority of the kidney but because the gills can transfer salt from the blood into the sea water that bathes them with virtually no loss of water. In this operation the animal is, in effect, concentrating the sea itself, an operation that requires physiological work and the expenditure of valuable energy in proportion to the quantity of salt thus transferred and the quantity of water gained in the body. Consequently the formation of urine is reduced to a very low level, generally to the minimal level set by the quantity of metabolic products requiring excretion. The common eel, *Anguilla*, for example, which can live in both fresh and salt water, has a urine flow in fresh water ranging up to 150 cc. per kg. per day, and this urine is very dilute; but when in salt water, the urine flow is generally less than 5 cc. per kg. per day and the urine is as concentrated as it can be made within the limitations of the fish kidney. These figures are roughly typical of all fresh- and salt-water fishes that have been studied.

This drastic requirement for water conservation in the marine teleost has had the most marked consequences on the structure and function of the kidney of any event in vertebrate history. With no excess water to be excreted, extravagant glomerular filtration is disadvantageous because every volume of filtrate formed requires the extensive reabsorption of both salt and water in the tubules. Hence the filtration rate is reduced to minimal levels by what appears to be an active inhibition of glomerular activity, probably by constriction of the glomerular arterioles. To compare the fresh-water catfish *Ameiurus nebulosus* with the marine sculpin *Myoxocephalus octodecimspinosus*, for which fairly accurate data are available: the filtration rate in the former ranges from 120 to 200 cc. per kg. per day; in the latter, the lower (and probably normal) values fall below 12 cc. per kg. per day, a reduction in glomerular activity of ten or more to one. Under these circumstances the glomeruli become more or less superfluous, and the stage is set for natural selection to favor those mutant forms that have intrinsically the *poorest* glomeruli, and hence for the evolution of an aglomerular kidney in which urine formation depends entirely on the excretory activity of the renal tubules. And so it has come about that among the permanently established marine teleosts we find all degrees of glomerular degeneration, culminating, in certain families, in wholly aglomerular kidneys in which the tubules begin (or end) blindly like a finger cot.

The discovery of the aglomerular fish kidney by A. Huot in 1897, since confirmed by many investigators, came as a surprise to renal physiologists who had been familiar only with the structure of the nephron in the higher animals, in the fresh-water fishes, and in the Amphibia. The aglomerular fishes were admittedly queer animals out of the sea—first the goosefish or angler, *Lophius*, then the toadfish, *Opsanus*, the sea horse, *Hippocampus*, and the pipefish, *Syngnathus*—one after another these marine fishes came under investigation to refute the assumption that glomeruli are basically necessary for urine formation, and to demonstrate that the tubules can excrete substances as well as reabsorb them.

The question of whether tubular excretion was even possible had been a subject of heated controversy in renal physiology for nearly a century. At one extreme, some investigators held that the glomerulus is merely a device for regulating the blood flow through the kidney, and that all the important constituents of the urine are excreted by the tubules. At the other extreme were those who contended that all the urinary constituents are separated from the blood by filtration through the glomeruli, and that the function of the tubules is limited to the reabsorption of water and other valuable constituents from the glomerular filtrate.

Now came the studies of Marshall and A. L. Grafflin, of Edwards and others, on the goosefish and other aglomerular forms, that demonstrated that the aglomerular fish kidney could excrete nearly all the important constituents of the urine, such as creatinine, creatine, uric acid, trimethylamine oxide (a compound excreted in substantial amounts only by the fishes), magnesium, calcium, potassium, and sulfate; and, among foreign substances, iodide, nitrate, thiosulfate, thiocyanate, and many dyes and synthetic compounds which are entirely foreign to the body and which the renal tubules had never encountered in their revolutionary history.

The demonstration that the aglomerular fish kidney could excrete nearly all the important constituents of the urine at least established one broad qualitative fact: tubular excretion, in principle, was a reality. But to prove the existence of tubular excretion in the aglomerular fishes was only to prove the possibility of tubular excretion in other animals; the demonstration really answered no questions at all so far as the glomerular kidney of the other fishes, or the frog, or the dog, or man was concerned, for conceivably the situation might be different in every species, and certainly it would differ for different substances. To answer this question in any species required an adequate method of measuring the filtration rate, and the aglomerular kidney served to focus the attention of investigators on this problem.

But the aglomerular kidney also supplied a convenient test organ for ascertaining what types of compounds can be excreted by the tubules—at least in the fishes, and, inferentially, if in the fishes, then possibly in other animals—and what types cannot. Most notable among the substances which the aglomerular kidney cannot excrete are carbohydrates such as glucose, xylose, lactose and sucrose: as valuable foodstuffs, glucose and related carbohydrates have always been conserved by the organism, and it has never been necessary for the renal tubules to excrete them into the urine. It was this observation that in part supplied the lead from which inulin (also a carbohydrate) came to be used for measuring the filtration rate. Nor can the aglomerular kidney excrete protein: protein appears in the urine only in consequence of excretion through the glomeruli.

The aglomerular kidney forced the historically minded physiologist to review the over-all history of the kidney, to ask how come and why. By putting this broad question to nature he acquired a better view of the evolution of renal function, and hence a better theory of renal function. And from the synthesis of this new knowledge, which was the work of many men, came new methods which now with confidence can measure, with an error of only a few per cent, the filtration rate, the renal blood flow, and a multiplicity of tubular operations, in a patient lying comfortably in bed.

In addition to the partial or complete degeneration of the glomeruli, the kidneys of the marine fishes show other evolutionary changes traceable to a salt-water habitat. The glomerulus, it will be recalled, was evolved as a high-pressure filtering device, and to this end it was supplied with arterial blood from the renal artery, this blood being introduced into the glomerular capillaries at a relatively high pressure in order to effect filtration. As the blood flow through the glomeruli is reduced in volume, the blood supplied to the tubules by the postglomerular circulation must be reduced proportionately, and the renal-portal system becomes of increasing importance in sustaining the function of the tubules. Tubular excretion neither involves nor requires the filtration of any water: the solutes to be excreted pass from the peritubular capillaries to the excretory tubule cells by simple diffusion, probably carrying water with them by osmosis, and the renal-portal blood meets all requirements in a marine fish in which the filtration rate is markedly curtailed.

And as the glomerulus degenerates, in the evolutionary sense, the arterial circulation to the glomeruli is also obliterated, leaving the tubules supplied only by renal-portal blood, which enters the peritubular capillaries at a pressure too low to effect any filtration whatever. We conceive that the renal-portal system, derived from the protovertebrate, has always been valuable to the fresh-water fishes, since it is preserved in all recent forms, as well as in the Amphibia; but its value to the marine teleosts is beyond question. Without it, it is doubtful if there would be any fishes (apart from the elasmobranchs) living in the sea today.

The renal tubule has also undergone extensive changes in the marine teleosts. We have spoken of the tubule in fresh-water forms as consisting typically of two segments, a proximal and distal segment. In such marine fishes as have been studied in this respect, only the proximal tubule is present—the distal tubule has disappeared entirely. It is believed that among the functions attributable to the distal tubule are the excretion of a dilute urine (water diuresis) and the regulation of the acidity of the urine. Typically, however, the marine fish never goes into fresh water and has no excess water to excrete. And the acidity of the urine is fixed on the extreme acid side: were the urine to become neutral or alkaline, magnesium, of which substantial quantities are absorbed from sea water in the gastrointestinal tract, would precipitate as magnesium oxide or magnesium phosphate and clog the tubules and collecting ducts. It is a matter of inference that the disappearance of the distal tubule is related to one or the other, or both, of these circumscriptions, which render this segment useless to a fish living in salt water.

In the glomerular nephron of both the fresh- and salt-

water fishes, the proximal tubule is divisible into two
parts which can be distinguished by the size and structure of the cells. In the aglomerular nephron, the first
part, immediately following the glomerulus, has disappeared, leaving only the second part, so that the tubule
in the aglomerular fish is reduced to a single segment,
homogenous in structure and function throughout its
length, and capable of carrying out all excretory operations. Again, it is a matter of inference that the first part
of the proximal tubule in the glomerular nephron may
be specifically related to some reabsorptive function, and
that with the disappearance of the glomeruli this portion
proved to be superfluous and was discarded.

If one examines the 'evolutionary tree' of the recent
fishes it is found that in the primitive or centrally placed
families that exclusively inhabit fresh water, or have species in both fresh and salt water, or can migrate freely
from one habitat to the other, the glomeruli are relatively large and well developed (lungfish, bowfin, carp,
goldfish, sucker, catfish, silver perch, bass, sunfish, trout,
pickerel, and eel). In relatively unspecialized but permanently marine forms the glomeruli are reduced in size,
but they still have the appearance of fairly good function
(sculpin, sea bass, grunt, squirrelfish, haddock, sea
raven, boxfish, porgy, spadefish, triggerfish, and rose-
fish). In a third category the glomeruli are small and
poorly vascularized, and functional activity is probably
greatly reduced (sergeant major, cod, needlefish, billfish,
flying fishes, anchovy, brown tang, puffer, shellfish, and
blue-striped grunt). In the fourth category the glomeruli
have disappeared entirely, leaving a purely tubular kidney. The extreme aglomerular condition is not characteristic of any one phyletic group but occurs in several
wholly unrelated groups, though generally in forms
which the ichthyologist considers to be highly specialized—in other words among families distantly removed
from the primitive fresh-water stem—such as the goose-

fish (*Lophius*) of the North Atlantic coast; the toadfish (*Opsanus*) of the Atlantic coast from Brazil to Cape Cod, including the sheltered waters of the Chesapeake Bay and Long Island Sound; the midshipman (*Porichthys*) of the Pacific coast; the batfish (*Ogcocephalus*) of the West Indies; the marbled angler or mousefish of the Sargasso Sea (*Histrio*); the well-known sea horse (*Hippocampus*) and the pipefish (*Syngnathus*); the clingfishes (*Lepadogaster* and *Gobiesox*); and several exotic deep-sea forms.

And yet the picture is sometimes muddled: for example, the common longhorn sculpin (*Myoxocephalus octodecimspinosus*) invariably shows good glomerular development, while the closely related shorthorn (daddy) sculpin (*M. scorpius*) presents the unique situation that some individuals show good glomerular development whereas others, even under the most rigorous functional tests, are proved to be completely aglomerular. Grafflin believed that in *M. scorpius* the glomeruli were lost as the fish grew older and larger, but the more recent observations of R. P. Forster show no correlation with size or presumably with age, and Forster concludes that the aglomerular condition is randomly distributed within this single species.

The list of aglomerular fishes is not longer than it is chiefly because the highly specialized fishes from the ocean depths have not been carefully studied with respect to renal anatomy. The writer once labored under the delusion that he could tell an aglomerular fish by looking at the creature: in this he was quite wrong because there are many queer-looking fishes that still have fairly good glomeruli, but the intuition was not wholly false because the deep-sea fishes are among the strangest-looking of all animals; they are specialized for life in the oceanic depths into which scarcely a ray of daylight penetrates and where they are never disturbed by any marked change in temperature, or by rain, wind, or lunar tide (whatever submarine tides may sweep them

up and down). It is required of them only that they
swim about in their crepuscular and static world waving
their phosphorescent lures to guide unwary victims into
their gaping maws; they epitomize life living itself out
in the nearest approach to absolute quietude, and they
are correspondingly delicate and fragile things. Though
geologically recent in origin, they are, in respect to specialization, farthest removed from the primitive fresh-
water forms.

None of the elasmobranchs, in spite of their long residence in the sea, is aglomerular, having always had adequate water available for filtration, they have had no
need to abandon the glomeruli.

How long the marine teleosts have lived in the sea is,
as we have noted, an unsettled question, but we believe
that the structure of the kidney affords some information
on this point. The primitive 'ganoid' fishes—the gar pike
(*Lepidosteus*), the bowfin (*Amia*), the bichir of the
Nile (*Polypterus*), the sturgeon (*Acipenser*), the paddlefish (*Polyodon*) of the Mississippi—are either fresh-
water forms or migrate into fresh water at some period
in their life cycle, a statement that is equally true of the
'centrally placed' teleosts. With rare and obvious exceptions, all permanently fresh-water fishes are phyletically
primitive. And all primitive, fresh-water fishes have
large and well-vascularized glomeruli. The supposition
that these fishes have recently invaded fresh water from
the sea is incompatible with the structure of their
glomeruli: had they had a long marine history one
would expect them to show the reduction in glomerular
development characteristic of the more specialized marine forms, and we believe that, on the evidence afforded
by the kidney, we may infer that the 'ganoids' and
primitive fresh-water teleosts have had a continuous
fresh-water (or migratory) history since their origin in
the Devonian.

Conversely, the marine fishes are under continuous

selection pressure to reduce or even to obliterate the glomeruli, as so many of the highly specialized forms have done. If they have all had a continuous marine history of approximately equal length, and of as long a duration as, say, from the Carboniferous, why have the majority retained fair to good glomeruli? It is in better agreement with the evidence to suppose that our present marine forms have been established as permanent residents of salt water only during Cenozoic time, an interval possibly too short to permit universal adaptation to the physiological stresses of a marine habitat.

A few aglomerular fishes have secondarily left the sea to reinvade brackish or fresh water—the toadfish (*Opsanus*) of the North Atlantic coast (closely related to the typically deep-sea Lophiidae) enters the tidal creeks of the Chesapeake Bay, while the pipefishes (Syngnathidae) are represented by several species that live and breed in the fresh-water rivers of Siam, Malaya, the Philippines, Panama, and the West Indies. The toadfish and the East Indian fresh-water pipefish, *Microphis boaja*, however, remain aglomerular—it is not in the nature of evolution that the clock can be turned back to give them glomeruli again. But how this or other forms maintain water balance in fresh water has not been studied.

Some glomerular fishes, such as the killifish (*Fundulus*), the stickleback (*Gastrosteus*), and the common eel (*Anguilla*), can tolerate rapid transfer from fresh to salt water, quickly making the necessary adjustments in water excretion. It is an oft-told tale how the common eel breeds in the Sargasso Sea in the mid-Atlantic and how the young elvers return to fresh water in Europe and America where they reach maturity and spend five or six years before returning to the sea to spawn. In salt water the eel drinks sea water, but in fresh water the only water to enter the gastrointestinal tract is apparently accidental, though other fresh-water fishes, such as the goldfish, may ingest water when feeding on microscopic food. The sturgeon (*Acipenser sp.*) and the salmon (*Salmo salar*) spawn in fresh water, but grow to maturity in the sea; while the fresh-water trout, which are small members of the genus *Salmo*, frequently descend from the rivers into the coastal waters.

Both the eel and salmon make their migration from fresh to salt water with but little change in the osmotic pressure of the blood. That a reduction in filtration rate plays an important part is certain, but no data are available on any of these fishes during migration. Nor is anything known about how the marine teleost goes about reducing the filtration rate to levels that will enable it to remain in water balance. One bit of evidence indicates that this reduction is attributable to active constriction of the glomerular arterioles: namely, when marine fishes are kept in captivity under not too favorable conditions they ultimately sicken and die, and in this physiological debacle the filtration rate and urine flow increase to astonishingly high values as though some renal inhibitory mechanism had broken down and the animal was drinking itself to death in an effort to compensate for water loss. In the longhorn sculpin, for example, during this 'laboratory diuresis' (which is of unknown origin) the filtration rate may increase from 12 to nearly 200 cc., the urine flow from 5 to 100 cc. per kg per day. It has been suggested that loss of slime increases the osmotic theft of water through the skin, but the point is far from proved. Forster has shown that this 'laboratory diuresis' also occurs in the essentially aglomerular daddy sculpin, *M. scorpius,* in which the urine flow may increase from 2 to 40 cc. per kg. per day, as well as in the aglomerular goosefish, *Lophius.* This is not surprising because the basic fact is that if the fish drinks more sea water for any reason, it has more magnesium and sulfate to be excreted by the tubules, and increased excretion of these solutes will lead to an increase in urine flow. In any case, a salt-water fish is one of the most delicate experimental animals with which the renal physiologist has to deal, which is one reason why it is difficult to quote reliable figures on renal function in these forms.

Whatever the basic renal operations that permit some fishes to migrate from fresh into salt water, or from salt into fresh, it is appropriate to look on this migratory habit as an extreme specialization in itself, involving increased effectiveness of regulatory mechanisms common to both fresh- and salt-water forms. For the great majority of fishes, the physiological adjustments involved in even the slow transfer from one medium to the other are inadequate to permit survival, and they remain bound by physiological habitude to narrow ranges of salinity.

An exception to this statement is, however, observed in certain semitropical fresh-water pools that communicate sluggishly with the sea and that are located in areas rich in limestone, so that the fresh water has a high calcium and bicarbonate content. Hard-water pools of this nature are characteristic of the Andros Islands and northern Florida. In such pools many typically marine fish, such as snook, tarpon, and snappers, are to be found very much at home with fresh-water forms. It is generally not difficult for the paleontologist to distinguish marine fossil beds from those that are continental in origin; but where, in the geologic past, fresh and salt water have met under these conditions the faunal mixture may belie all other evidences and reduce his confidence to doubt. It would appear that either the high calcium content or the alkalinity of these hard waters makes the transition from salt to fresh water easier—but why, nobody knows. The question may seem trivial, but not to anyone who is concerned with the intricate problem of salt and water balance, with the proper functioning of the heart and blood vessels, with the blood supply to the brain, with the many problems that remain unsolved in respect to the regulation and distribution of the internal environment in man.

CHAPTER IX

THE REPTILES AND BIRDS

A commentator on earth's history must be cautious to avoid superlatives, and we have already spoken of the Pennsylvanian period in extravagant terms. Yet in passing to the Permian period an antithetic emphasis can scarcely be avoided, especially when even the cautious historical geologist calls the transition 'one of the great crises in the history of life.' The end of the Pennsylvanian was marked by the Appalachian revolution, which continued irregularly throughout the Permian, raising the definitive Appalachian Mountains from Newfoundland to Alabama and warping nearly all the Paleozoic formations into great folds—locally exhibited from the Blue Ridge in Virginia to the Alleghenies in Pennsylvania by remnants of mountains which, when young, must have rivaled the present Himalaya. It is estimated, for example, that the distance between Philadelphia and Altoona, Pennsylvania, now 177 miles, was shortened by a third or more simply by folding and faulting as the Appalachians were pushed into the air. The unstable Appalachian 'trough,' which ever since the Cambrian had subsided intermittently with every shudder of the earth and had been repeatedly the site of marine invasion, trapping some fifty thousand feet of sedimentary strata, was now raised so high that the sea has never invaded it again.

At the beginning of the Permian an inland sea, invading from the south, had occupied the Great Plains from Colorado to Ohio; then in the course of continental elevation this sea was drained to leave in Kansas and Oklahoma a dead sea which deposited deep beds of salt, and these in turn were overlaid with hundreds of feet of red mud, and finally with desert sands. Erosion leveled the older Colorado Mountains, depositing fourteen thousand feet of rich fossilized strata over the areas of Texas and New Mexico (now exposed in the Guadalupe Mountains); while as Utah, Nevada, Idaho, and Oregon were pushed from sea level into high relief, volcanic activity broke out in western North America for the first time in the Paleozoic era, penetrating granitic faults from central Mexico to northern California. This period saw the elevation of the Ural Mountains in Eurasia and other chains across southern England, Germany, and northern France, while much of the rest of Europe was covered for a long period by a great dead sea along the edges of which were sandy deserts.

In many areas of the world the Permian presents a picture of marked continental elevation bringing in its wake the inevitable climatic change. Between the early Pennsylvanian and the opening of the Permian the mean temperature in Europe and North America is estimated to have dropped from 53° to 32° F. The withdrawal of the great seas, which had covered so large a fraction of the Pennsylvanian continents, removed the thermal influence of equatorial currents and the stabilizing effects of large bodies of water, while the new mountain ranges interfered with the winds and increased precipitation on the windward side, decreasing it on the lee, so that deserts became more widespread than at any time up to the present. The southern ice cap crept north to the Tropic of Capricorn and glaciers spread at altitudes in South Africa, Australia, Argentina, southeastern Brazil, and India—there is even evidence of there having been a small glacier near Boston. The Permian ice-age was, however, of relatively short duration: the climate, taken through the whole of the period, was one simply of chilly aridity, both cold and dry.

With this drop in temperature the swamp-dwelling flora, so rich in the Pennsylvanian, contracted to isolated oases, the insects became smaller and more varied but less abundant, and the Amphibia, that had fed upon these insects and luxuriated in the Pennsylvanian swamps, faced not only a scarcity of food and an excess of chilliness and aridity, but another crisis that struck at the very roots of their existence—their mode of reproduction. The adult might fight for survival in dwindling pools and along seasonal streams by effort and cunning in catching spiders and dragonflies, but for the eggs and helpless young to be exposed in a highly restricted habitat to ravenous fishes searching out every edible morsel meant destruction. A frog lays many hundreds of eggs each season, most of which under favorable conditions may hatch into tadpoles with fish-tail and fish-gills, but these must live in the water for several weeks until they metamorphose into terrestrial-living adults that can escape their aquatic enemies, and the probabilities are great that in that interval all but a few will be devoured. The adage about big fishes eating little fishes is true enough, but the fish tribe survives because the female lays thousands of eggs—in some cases up to a million per season. Nowhere is nature as careless of the individual in order to preserve the race. The Amphibia, however, lay far fewer eggs and under the pressure of narrowly confined breeding pools the reproductive waste must have increased enormously in the Permian until selection favored those forms whose eggs could be deposited in isolated hiding places on the land. In the extreme, this meant that the egg must carry with it its own water supply, and it appears to have been in the amphibian suborder of the Embolomeri, and by forms akin to *Seymouria* (see Figure 8) that the reproductive pattern departed from the aquatic mode to produce the reptiles. As

Romer says, 'it was the egg which came ashore first; the adult followed later.'

The first great invention of the 'reptiles' (we use the word with reservations at this time) was the 'amniotic' egg, so-named because it contains a closed sac or 'amnion' which is filled with fluid and which supplies an artificial aquatic environment in which the embryo can develop through the larval stage into a miniature but adult animal. (Figure 9.) A second membrane, the 'allantois,' creates a cavity (the allantoic cavity) which communicates with the embryonic kidney and serves as a repository for such urine as is formed. A third membrane, the 'chorion,' initially lining the eggshell, unites with the allantoic membrane to form a respiratory mantle, richly perfused by blood vessels and serving to take in oxygen and give off carbon dioxide for the developing embryo. A yolk, greatly enlarged beyond that of the fishes or Amphibia, supplies food for sustained development. Such is a hen's egg, which is not significantly different from that of the reptiles. The quantity of yolk in the eggs of the Amphibia and of the fishes varies widely from one species to another, and the enlargement of the yolk sac in the amniotic egg represents merely a quantitative variation, while the chorion can be considered to be an elaboration of the primitive egg membrane. The amniotic membrane, however, with its artificial internal environment for the embryo, is a novel invention which has the appearance of a most happy accident.

These three embryonic patents, the amniotic, the allantoic, and the chorionic membranes, are practically identical in structure and function in the birds and reptiles and persist with greatly elaborated function in the mammals, the amnion here supplying the fetal membrane filled with amniotic fluid, the chorion becoming the fetal contribution to the placenta, the allantois part of the umbilical cord. In the reptiles, birds and two primitive mammals—the duckbill, platypus, and the spiny anteater, echidna—the egg is covered with a thickened shell which may or may not be calcified, but it always remains porous enough to permit the passage of oxygen and carbon dioxide. This shell is, however, relatively impermeable to water, thus preserving the water supply of the amnion. Where the Amphibia had relied on external fertilization of the egg, this cleidoic (*kleistos* = closed) egg had to be fertilized in the female before it was covered with its shell, an operation that was effected primitively by the intromission of sperm into the cloaca without the aid of special sexual organs.

Because of the basic identity of pattern in the amniotic egg—which is one of the most impressive features in vertebrate history—the reptiles, the birds (which are of reptilian origin), and the mammals are frequently if informally called the 'amniotes.' It is clear that the amniotic egg (the oldest known fossilized specimen comes from the Lower Permian) must have been invented before the ancestral lines, which led, respectively, to the reptiles and the mammals, had diverged—perhaps by a form akin to *Seymouria* (see Figure 8) in the late Pennsylvanian when the Appalachian revolution was bringing aridity upon the land—and it is equally clear that the first 'amniote' was no more a reptile than it was a mammal. The great physiological differences between the reptiles and mammals, differences which separate them as markedly from each other as from the Amphibia, warrant the not entirely frivolous suggestion that a new class (or superclass) be recognized, intermediate between the reptiles and mammals on the one hand and their amphibian root on the other, and perhaps to be called the Amniota (Pennsylvanian to Permian; families, genera, and species *incertus*), of which no relict survives. In the absence of the paleontologist's blessing, we shall simply refer to these intermediate forms as the 'nascent' amniotes.

In the early Permian, this nascent amniote stock—perhaps all of fifty million years of age—separated into several branches of which one was to lead to the great Jurassic and Cretaceous reptiles and, through a side branch, to the birds; while another branch was to lead by way of the Permian Theriodontia (*therion* = beast; *odon* = tooth) and the Triassic Cynodontia (*kyon* = dog; *odon* = tooth), to the mammals. Here, for the moment, we follow the evolution of the reptiles and the birds. (Figure 9)

The oft-quoted statement that the birds are only glorified reptiles that have gained wings and lost their teeth sums up the many basic affinities between these two classes. The birds and mammals, on the other hand, are together distinguished from all other vertebrates by their warm-blooded or homeothermic (*homoios* = like or similar; *therme* = heat) state, the body temperature being maintained by metabolic and circulatory adjustments in the range of 97° to 103° F.; while the reptiles (like the Amphibia and fishes) are cold-blooded or poikilothermic (*poikilos* = many-colored; *therme* = heat), the body temperature being without significant control. At first sight it would seem that so notable a feature as warm-bloodedness must connect the birds and mammals in a common origin, but the paleontological and physiological evidence is overwhelmingly against this supposition; rather it must be accepted that warm-bloodedness has been evolved independently in the two classes: their ancestral lines had separated by the Permian (if not earlier) and at a time when the nascent amniote was still cold-blooded.

The birds (or a bird) first appear in the fossil record of the Upper Jurassic, in the fine-grained lithographic sandstone (such as that from which Currier and Ives and other nineteenth-century lithographers made their prints) mined in the quarries at Solenhofen, Bavaria. Only two such fossil birds (and one feather) are known, and though they closely resemble each other they are usually considered as representing different genera, *Archaeopteryx* (*archaios* = ancient; *pteron* = wing), and *Archaeornis* (*archaios* = ancient; *ornis* = bird). *Archaeopteryx* has been called the ideal fossil because it so perfectly connects the birds with their reptilian ancestor. Whether one calls *Archaeopteryx* a reptile or bird makes little difference. It possessed a long reptilian tail composed of twenty lizard-like vertebrae with feathers arranged in pairs along the axis rather than fanwise (as in the modern birds); the 'wings' ended in 'hands' with a thumb and three separate metacarpal bones each carrying a free finger and each ending in a reptilian claw; and the upper and lower jaws were lined with reptilian teeth set in individual sockets. The hind legs were, however, avian, and the clawed feet were adapted to arboreal life. That *Archaeopteryx* could at least glide from tree to tree is beyond question, and paleontologists call it a bird and give it a subclass all to itself, partly because of its anatomical affiliations and partly because of its feathers. By this taxonomic identification they imply that it was warm-blooded. If this supposition is correct, it places the beginning of the evolution of warm-bloodedness in the avian stem in the Jurassic, possibly even in the Triassic period. But despite the fact that they are warm-blooded, the birds are typically reptilian, especially in matters concerning water balance.

It was one thing for the Pennsylvanian-Permian amniotes to devise an egg with its own internal environment to sustain the biochemically helpless embryo; it was another for the adult to solve the problem of living under arid conditions on a minimal quantity of water. The water-permeable skin of the Amphibia had to be replaced with an impermeable hide: the skins of the early reptiles are poorly preserved in the fossil record but doubtless most of them were variously covered from snout to tail with small scales, scutes, or plates derived from the thickened, cornified epidermis in a manner foreshadowed among some of the Pennsylvania Amphibia and persisting primitively in the warts of living toads. But waterproofing of the body did not wholly solve the problem because, with neither gills nor skin to participate in the process, the kidney now became the sole means of regulating the composition of the body fluids and the water requirement for urine formation set the limits to the animal's freedom and independence of environment. Now, more than ever before, survival hinged on the efficiency of the kidney in maintaining the water balance of the body.

This crisis was met by a profound transformation in the function of the reptilian kidney, supplemented by a change in protein metabolism. It has been noted that the major waste product requiring excretion by the kidney is the nonvolatile end-product of the metabolism of protein nitrogen. In the fishes and Amphibia this had been urea, and so it continued to be in the 'amniotic' stem that led to the mammals. The reptiles, however, faced with a need of water conservation no less pressing than that encountered by the marine elasmobranchs and teleosts, overhauled their method of protein metabolism and replaced urea by uric acid, which, molecule for molecule, carries twice as much nitrogen out of the body, and thus osmotically requires only half as much water for the same quantity of protein metabolized But more important is the fact that uric acid is relatively insoluble in water, and yet it readily forms a highly supersaturated solution from which it separates as a fine, amorphous precipitate or as microscopic crystals. Once out of solution, it exerts no osmotic pressure. The reptiles (and birds) take the utmost advantage of this unique property: they deposit uric acid by tubular excretion (in addition to filtration) in the tubular urine in very high concentrations—the ureteral urine of the chicken, for example, may contain the extraordinary concentration of 21 per cent, or some three thousand times the concentration in the blood—and as this urine passes to the cloaca the greater part of the uric acid precipitates out of solution and leaves practically all the water osmotically free to be reabsorbed. This reabsorption is carried out by regurgitation of the urine into a terminal, specialized segment of the intestinal tract, and the semisolid residue of uric acid is then defecated as a white paste with the intestinal residues.

At the typical osmotic pressure of amphibian or reptilian blood (which is about the same as that of the bony fishes and the mammals), the 320 mg. of urea formed from one gram of protein would require 20 cc. of water for its excretion in an isosmotic urine—that is, one having the same osmotic concentration as the blood; in the reptiles and birds, where protein nitrogen is metabolized to uric acid, the final 'urine' as defecated from the cloaca may take the form of an almost dry concretion, the nitrogen derived from one gram of protein requiring less than 1.0 cc. of water, and possibly as little as 0.5 cc., for its excretion, a tremendous saving in water. Since the oxidation of one gram of protein itself supplies 0.4 cc. of water, the animal is placed on an almost self-sustaining basis.

One might expect the reptiles and birds to be aglomerular, and they are nearly so, but not quite. All that have been studied have small glomeruli in which the capillaries are reduced to but two or three short loops, very different from the large, well-vascularized glomeruli of the Amphibia. The reason that they have not forfeited their glomeruli is perhaps that the tubules are unable to excrete sodium chloride (and possibly other salts), and they must rely solely on the filtering bed for salt excretion. Moreover, in so far as the uric acid precipitates out of the tubular urine it no longer automatically draws water with it (as do the salts in the urine of the marine fishes) and it is questionable whether water could be made available by the tubules to wash the uric acid suspension into the cloaca. Consequently, in reducing the glomerular filtrate to a feeble stream they have possibly gone as far toward the aglomerular condition as they can.

The uricotelic (uric acid-excreting) habitus is characteristic of all birds, so far as is known, and of all reptiles studied that have persisted in an arid habitat, and we take it to be a fundamental biochemical characteristic of the reptilian-avian stock, and one that was evolved before the separation of the avian from the reptilian stem. The uricotelic habitus is not present in any known amphibian or mammal, all of which excrete urea (except for small quantities of uric acid formed in the specific metabolism of nucleoprotein), and it must have been invented by the reptilian branch of the nascent amniotes at the time when they were undergoing adaptation to terrestrial life, during late Pennsylvanian or early Permian times.

No other vertebrate class has undergone such spectacular evolution with such rapidity as have the reptiles. Before the end of the Permian they were represented by several highly divergent orders, and in the Triassic they grew in size if not in variety, spawning the first dinosaurs (*deinos* = terrible; *sauros* = lizard). These dinosaurs were small in comparison with those of the later Jurassic and Cretaceous—only few reached a length of more than 10 to 15 feet—but they all were bipedal and were adapted for running on their hind legs and used the forelegs for grasping vegetation or for digging. The long, stout tail was used, as by the kangaroo, for balancing and bracing the body and, again, like that animal, they held the tail off the ground when running. Footprints are more abundant than skeletons, because the Triassic red beds afforded a poor environment for preservation, but many Triassic mud-flats reveal the criss-cross trails of running dinosaurs, the mud frequently spattered by raindrops that fell just before or just after the visitation. Two groups of Triassic reptiles took to the sea as a permanent abode: the dolphinlike ichthyosaurs and the turtlelike plesiosaurs, both probably having invaded salt water by way of the marshes.

The Jurassic saw the tendency to increased size culminate in the truly gigantic dinosaurs The American remains of these animals have all come from the so-called 'Morrison formation' of the Upper Jurassic, named for its excellent exposure near that Colorado town, but the formation extends over an area of a hundred thousand square miles of the Rocky Mountain region. This is a typical terrestrial formation representing a great alluvial plain crossed by sluggish streams flowing eastward and interrupted by swamps and small lakes. Though, as in all other times, the climate varied considerably from one locality to another, the Morrison beds reveal a warm temperature and high humidity, supporting lush vegetation of scouring rushes, tree ferns, cycads, ginkgos, and conifers—all evergreens, for the deciduous hardwood trees were as yet unknown. The Jurassic flora was much the same all over the world, in North America, Siberia, arctic Alaska, Spitsbergen, Scandinavia, England, and even in Louis Philippe Peninsula below Cape Horn. In this favorable milieu the reptiles developed to their fantastic and now familiar extremes. The pterodactyls had extended the little finger into a long arc from which a leathery wing stretched along the arm to the body, they certainly were capable of soaring and some may have been capable of sustained flight, though on the ground they walked on all four feet. They were not birds and not related to birds, but cold-blooded flying reptiles and more properly likened to winged dragons.

Among the dinosaurs, one of the best-known American forms, *Brontosaurus*, reached a length of 65 feet and a weight of some 30 tons; the more slender *Diplodocus*, a length of nearly 80 feet. *Brachiosaurus*, with giraffelike proportions and long forelegs, carried its head 34 feet above the ground. *Stegosaurus* was covered with armor fabricated of massive plates and weighed an estimated 10 tons. These were primarily herbivorous animals, but the somewhat smaller *Allosaurus,* which had a length of only 30 feet, and the little *Compsognathus,* only 2½ feet long, had developed carnivorous habits and probably preyed upon their less agile fellows. In well-preserved fossils of Jurassic ichthyosaurs, unborn young have been found inside the body, showing that as an adaptation to their marine habitat they had become ovoviviparous, a notable adaptation since all recent aquatic reptiles come ashore to lay their eggs. (Rattlesnakes and a few other recent reptiles are ovoviviparous, but not in adaptation to a marine habitat.)

In the Cretaceous, when reptilian evolution reached its peak, *Tyrannosaurus rex,* the mightiest flesh-eater ever to live on earth, stretched 47 feet from nose to tip of tail and, erect on his massive hind legs, his head towered 20 feet above the ground. The duckbilled dinosaur, *Trachodon mirabilis,* was almost as large, while some of the *Ceratopsia* or horned dinosaurs were twice as bulky as the greatest living rhinoceros. In the sea some of the plesiosaurs reached a length of 40 to 50 feet, and competed with marine turtles 11 feet long and with a newly evolved group, the mosasaurs, which had taken to carnivorous piracy in the water. The pterodactyls, though less numerous than in the Jurassic, also went in for size: *Pteranodon,* for example, had a wingspread of 25 feet, exceeding any other winged creature of all time and truly meriting the title of winged dragon. It is believed that this creature, like the flying reptiles of the Jurassic, lived on fish, which it possibly caught in the salt marshes by diving from high cliffs, but how it got on to the cliffs and how it got out of the water remain a mystery.

The transition from the Mesozoic to the Cenozoic has been called 'the time of the great dying.' Every one of the terrestrial and aquatic dinosaurs went down to extinction, leaving of the great Age of Reptiles only the turtles, snakes, lizards, crocodilians, and the primitive lizard, *Sphenodon,* to survive in the modern world. No class was spared, for even the fresh-water and marine fishes suffered decimation, while among the ammonites and belemnites, and the reef-forming clams of the ocean floor, not a family survived to see the dawn of the Cenozoic era.

There is little about the end of the Mesozoic to explain this world-wide destruction of many orders and innumerable species of animals. The coal beds of New Mexico, Colorado, Utah, Wyoming, and Montana were laid down in the Upper Cretaceous (at the end of the Mesozoic) and on into the Eocene (at the beginning of the Cenozoic) without a marked hiatus; the oil beds of eastern Texas are of Upper Cretaceous origin and, late in this period, figs, breadfruits, cinnamons, laurel, and tree ferns were growing in Greenland, while cycads, palms, and figs were growing in Alaska. There was certainly no great icecap in the Northern Hemisphere at the close of the Mesozoic, and only local glaciation of elevated areas occurred in the Eocene. In short, there was no such catastrophic change in climate as marked the Permian 'stricture,' when frigidity spread over so much of the world to blank out a large fraction of Paleozoic life. The 'time of the great dying' does, in fact, coincide with another shudder of the earth—the Laramide revolution—and a warping of the continents which again brought in its wake withdrawal of the seas, a drop in temperature, even if moderate, and contraction of the swampy lowlands with a marked change in vegetation all over the world; but the dramatic thing is that these rather undramatic changes, when added together, should have spelled death to so many Mesozoic forms.

It is futile to say that the great reptiles failed in the evolutionary sense, because too many other animals failed with them in this mysterious crisis. On the contrary, the reptiles did not fail, because they established themselves as the first complete masters of the land. Many of the Permian and later forms give the appearance of being amphibious and water-bound, but in view of the nature of the reptilian egg it is probable that even these bred away from the water, and laid their eggs in warm crannies in the sand. One may assume that many of the Mesozoic reptiles may have spent the greater part of their lives away from watercourses, as do the arid-living reptiles today, returning to the lush valleys as a matter of convenience rather than necessity. As judged by the surviving members of the order, most of the terrestrial forms were able to live almost independently of a free water supply by virtue of their amniotic egg, their impermeable skin, and their uric acid habitus. The reptiles were deficient in only one notable respect—brains. If *Tyrannosaurus rex* had the smallest brain per unit of body weight of any vertebrate, it did not need a larger brain in order to survive; it was so well adapted to its world that lack of cunning did not count heavily against it. But it was otherwise with the small and poorly adapted mammals that *Tyrannosaurus* unwittingly trampled underfoot: in these mammals the evolution of the brain was to become of paramount importance.

CHAPTER X

THE MAMMALS

Equaling if not transcending in importance the chemical composition of the internal environment is the regulation of the temperature of that environment to a relatively fixed range, so that, in truth, as Bernard said, the animal lives in a sort of 'hothouse,' protected from the thermal vagaries of the outside world. In the entire animal kingdom this temperature regulation has been achieved twice only, in the birds and in the mammals.

The time and circumstances of the evolution of warm-bloodedness in either of these classes have not been widely discussed by the paleontologists for the obvious reason that the fossil record affords no direct evidence on the matter. Mammals are insulated against heat loss by means of hair, which is, like feathers, an outgrowth from the epidermis, though wholly different in development and structure; but unlike feathers, hair is not easily fossilized and it is not known to what extent it was present in the early forms, which are known only from their jaws, teeth, and skeletal remains. The fossil record with respect to the earliest birds, on the other hand, is limited to two specimens (*Archaeopteryx* and *Archaeornis*) from the Jurassic. Consequently we can only speculate as to when either class became warm-blooded.

Despite many gaps, the evolution of the mammals from the reptilian stem is one of the better known chapters in paleontology. The definitive mammalian tendency in respect to skull and teeth is first evident in the Pelycosauria (*pelykos* = basin; *sauros* = lizard, so named in reference to the form of the pelvis) of the Pennsylvanian, indicating that the separation of the mammals from the reptiles followed shortly after the evolution of the amniotic egg. More distinctively mammal-like forms appear in the Permian Therapsida (*therion* = beast; *apsides* = arch, so named because the temporal region of the skull is like that in the mammal), which are known by two hundred or more genera and an unknown number of species, thousands of museum specimens and 'countless thousands still in the rock.' It has been suggested, but not proved, that the advanced therapsids, as represented by the Theriodontia, may have been warm-blooded and may have nursed their young. In the Cynodontia of the Triassic the teeth had begun to acquire the characteristic mammalian form of incisors, canines, premolars, and molars, the roots of which were implanted in sockets. Here mammalian evolution divided into two streams, one to produce the Monotremata or egg-laying 'mammals'— Simpson calls these 'highly modified surviving therapsid reptiles, mammals by definition rather than by ancestry' —of which the duckbill (platypus), and the spiny anteater (echidna), of Australia, are the only surviving members, both so primitive that they lay shelled eggs like their reptilian ancestors.

The platypus is, in fact, one of nature's strangest mixtures and illustrates something of the transition from the reptilian to the mammalian level. The typical mammal is, by definition, a warm-blooded animal whose body is covered with hair, and whose young mature within the uterus and are born alive and are nursed from milk glands bearing nipples. The platypus, though covered with sleek mammalian hair resembling sealskin, lays tough-skinned, reptilianlike eggs which it incubates with its body, but the young are immature at birth (as with many mammals) and nurse from a rudimentary mammalian milk gland which lacks a nipple. It has a flattened, ducklike bill, and teeth are present only in the young as vestigial molars. It is in no way related to the birds, in spite of its bill and the fact that its feet are webbed like those of a duck, both characters being secondary. The bill is a soft, kidskinlike leathery expansion of the muffle of the marsupials, and the webbing between the toes is only an exaggeration of that between the human fingers. It is, in short, a grotesque mixture of reptilian and mammalian characters, with some specializations of its own. The platypus's five-toed hands and feet are well adapted to digging, and it lives in burrows along the banks of the streams and lakes of Australia and Tasmania, feeding on insect larvae, worms and aquatic plants. Its consumption of food in captivity is phenomenal: a four-pound male specimen in the New York Zoological Park consumed daily 20 to 30 live crayfish (more if available), one frog, a dish of duck-egg custard (1 to 1½ eggs), and one pound of live earthworms (and only difficulties of supply kept the worm consumption down to this level)—adding up to a gross weight of two and a half to three pounds. Nothing is known, regrettably, about nitrogen excretion in either the spiny anteater or the platypus, nor will the alert Zoological Park authorities allow an experimentally minded investigator within less than looking distance of the tremendously valuable animals in its platypusary.

The second stream of mammalian evolution stemming from the cynodonts led to the Jurassic Pantotheria (*panto* = all; *therion* = beast), from which were derived all later mammals, the most primitive of which are the Marsupialia, or pouched animals, which include the kangaroo and opossum (the latter the most primitive mammal in the northern world)—whose young are born alive but in such an immature state that they must be carried in a pouch by the mother for a considerable

period after birth. Also from the pantotherian stem, there
evolved the true or placental mammals, the Eutheria
(= true beasts), in which the young are (with few exceptions, including monkeys, apes, and man), physiologically capable of taking care of themselves as soon as
they are weaned. In the Eutheria, the amniotic egg is
left within the womb to mature into a miniature adult,
thus supplying the developing embryo with an ideal 'external environment' until such a time as it can manufacture its own internal environment by foraging for
food and water for itself.

The homeothermic mechanism is not perfected in all
mammals. In nearly all forms the regulation of body
temperature is poor in the newborn—a two-day-old
mouse, for example, is essentially a poikilothermic animal and does not suffer from the circumstance that its
body temperature drops considerably when the mother
leaves the nest. The platypus and the echidna, as well
as the opossum and other marsupials, the armadillo, bat,
woodchuck, hamster, sloth, and numerous others, are not
even strictly homeothermic as adults: during warm
weather they have a somewhat lower and more fluctuat-
ing body temperature than the advanced mammals, and
during cold weather they hibernate and the body temperature falls to a level only slightly above that of the
environment. Hence they can more properly be called
'heterothermic.' (Many truly homeothermic animals,
such as bears and foxes, also hibernate in the winter,
but without marked reduction in body temperature;
they simply go to sleep and maintain their metabolism
at a slightly reduced level by burning their rich stores of
body fat.) Newly hatched birds also have poor body-
temperature regulation, but no adult heterothermic bird
is known, possibly because selection has left us only a
sample that is highly proficient in this respect.

As homeothermy is not universal among existing mammals, it is probably wise to look upon the Jurassic mammals as at best heterothermic. Nevertheless, the tend-

ency to maintain body temperature and physical activity in cool weather must early have become an important factor in mammalian evolution and, if skulls and teeth are any criterion of 'mammalness,' selection may have been working in this direction even among the Permian Therapsida.

The Permian was characterized by climates more rigorous and sharply zoned than had prevailed since the pre-Cambrian, with widespread glaciation at altitudes and a marked drop in temperature in all latitudes. It was also a period of aridity, in which the great coal-making swamps of the Pennsylvanian contracted to isolated locations, to be replaced in large part by desert sand. So severe were the conditions of life that many types of animals, both on land and sea, went down to extinction, and others were greatly reduced in varieties and numbers—for which reason the paleontologist speaks of the 'Permian stricture' as though it were a needle's eye through which the rich and diversified fauna of the Pennsylvanian could not pass.

It is tempting to see the pressures of selection, in the form of low temperatures prevailing over so large a part of the world, and widespread aridity, working more or less independently on the derivatives of the cold-blooded amniotes which, in the Pennsylvanian, had already partially adapted to aridity through the evolution of the amniotic egg. (See Figure 9.) In the Permian the reptiles went one step further in adaptation to aridity by the evolution of the uric acid habitus, and it was possibly not until the late Triassic or early Jurassic that a bipedal stem of these uric acid-excreting reptiles adapted to frigidity, to give rise to the warm-blooded birds. There is nothing, however, to argue against the view that the pro-avian reptilian stem may have acquired some control of body temperature as far back as the Permian—no reptile that survives today can throw any light on the matter, the nearest relatives to the birds being the crocodiles and alligators.

Along the mammalian line, however, the Therapsids, retaining the urea habitus of the nascent amniotes, adapted to Permian frigidity to give rise to the hetero-thermic theriodonts and later mammalian forms.

It is scarcely worth arguing how much biological ad-vantage over the reptiles the birds derived from their warm-blooded state; mere preponderance of numbers in respect to families, genera, and species does not prove the case for biological advantage, any more than does the ability to soar in the air, and in the long view the birds flew into a blind alley when they took to flight. What they might have gained by their warm-blooded-ness they lost by abandoning their forelimbs to the role of wings. But that warm-bloodedness worked ultimately to the great advantage of the mammals is not to be denied: they were earthbound, it is true, but this cir-cumstance rather than handicapping them was ulti-mately to profit them—because, to satisfy the hunger that was now greatly enhanced and constantly sustained by the demands of body-temperature regulation, they were impelled to remain active for long periods through-out the year. When winter came the birds could fly to warmer zones and the reptiles could sink into the obliv-ion of cold sleep; but the mammals had to stick it out where they were and continue to eat and drink to main-tain the constancy of their internal environment. It was simply a case of keep active and awake—or else!

Throughout the Mesozoic the primitive mammals made little progress, possibly because grasses, cereals and fruits were still rare, and competition from the car-nivorous dinosaurs was keen. Their opportunity did not come until the Laramide revolution and the opening of the Cenozoic (*kainos* = recent; *zoe* = life) era, when they underwent rapid multiplication and transformation in modern forms.

The whole of the Cenozoic, from the end of the Cre-taceous to what the geologist calls 'Recent' time (which includes the present), has been a period of irregular continental elevation and mountain building to which may be attributed all the rugged profiles of the present world. In the Middle Cretaceous the Rocky Mountain area had been submerged in a marine trough 100 miles wide and extending from Alaska southward and eastward into the Gulf of Mexico; in the late Cretaceous, continental uplift had drained the salt water from this trough to leave swamps that made many a rich coal bed from Alberta to Mexico. Then in the Laramide revolution (named after the Laramide Range in Wyoming), which ushered in the Eocene, the North American continent buckled along this trough to give birth to the basal structures of the present Rocky Mountain System —over 3000 miles long and, at its maximum extent across eastern Colorado to central Idaho, 500 miles wide, with the Front Range at Longs Peak in Colorado reaching a height of perhaps five miles—the most severe disturbance to be experienced by this continent since pre-Cambrian times. Another range extended from Trinidad southward through Colombia to beyond Cape Horn, a distance of nearly 5000 miles. In Europe, the Eocene saw the first decided upthrust of the Alps, though these, like the Laramide Rockies and Andes, have subsequently suffered extensive erosion, and their present altitude is attributable to continued uplift in the Pliocene and Pleistocene—that is, within the last two million years or so.

The present Sierra Nevada are also relatively young mountains, and represent a colossal block 100 miles wide and 300 to 400 miles long, the eastern edge of which was tilted in the Pleistocene to an elevation of 13,000 feet, the western edge depressed perhaps 25,000 feet below sea level in what is now the 'great trough' of central California, between the Sierra and the Coast Range.

Also dating from the last geologic period are the Himalaya, the foothills of which have been lifted 6000 feet since the middle of the Pleistocene so that marine formations laid down in the Eocene, just after the dinosaurs had become extinct, now lie at an elevation of 20,000 feet. The Grand Canyon of the Colorado River, in places more than a mile deep, was etched in Pleistocene and Recent time when the land rose by that amount and more In short, the spectacular reliefs of the world to-day are all relatively new wrinkles in the earth's unstable crust and are largely related to the most recent episode of continental elevation, the Cascadian revolution, which derives its name from the Cascade Mountains around Puget Sound.

The elapsed time from the beginning of the Cenozoic era to the present is roughly 55 million years, or 10 per cent of the history of the vertebrates. Yet at the end of the Mesozoic the mammals were already nearly 100 million years old—two-thirds of their total history is obscured in the Mesozoic. For mysterious reasons they had to wait until the world was rid of the dinosaurs before they could come into their own. Someone has facetiously said (in the absence of a better reason) that perhaps the dinosaurs became extinct because the little mammals ate up all the dinosaur eggs. More seriously, as Simpson suggests, the facts better fit, though they do not prove, the reverse proposition. that in the late Cretaceous the birds and mammals replaced the reptiles because the reptiles had dwindled in numbers or had become extinct.

It is, however, cogent to note that the mammals are, as a class, almost as well adapted to life under arid conditions as are the reptiles, an adaptation that is achieved by virtue of the fact that the mammalian kidney can elaborate a urine that is substantially more concentrated, in respect to osmotic pressure, than the blood.

This concentrating power appears to have been evolved as a concomitant of the evolution of warm-bloodedness. To raise the body temperature above that of the environment, it was necessary not only to reduce heat loss through the skin by covering the body with fur, but also to establish control of the peripheral blood ves-

sels so that blood could be shunted away from the surface
when heat needed to be conserved. To keep the body
temperature down in hot weather and during physical
activity, these same blood vessels had to be dilated and
the skin wet with sweat, the evaporation of which would
increase heat loss. Thus a major factor in the evolution of
the homeothermic state was the increased development
of the nervous control of the blood vessels, both in the
skin and throughout the internal organs of the body,
one of the consequences of which was that the arterial
blood pressure was not only stabilized but set at a
relatively higher level than in antecedent forms. Elevation of body temperature also required an increased supply of fuelstuffs for the tissues, increased oxygen consumption, and increased production of carbon dioxide
and other metabolic products—all of which required an
increased rate in the circulation of the blood. But increased blood pressure and increased circulation meant
an increase in the rate of glomerular filtration, which in
turn required increased tubular reabsorption of valuable
constituents from the filtrate, and particularly an increased capacity to conserve the valuable water in that
filtrate by making a superconcentrated urine, a task
achieved in a truly significant manner by the mammals
for the first time in either vertebrate or invertebrate history. Since the mammals persisted in the urea-excreting
habitus, and since urea is a very soluble substance with
relatively great osmotic pressure per gram of protein
nitrogen, the homeothermic state invited them to go in
the direction of a superconcentrating kidney. The birds
exhibit this concentrating power to a slight degree, but
they were spared the requirement of an osmotically concentrated urine because of their uric acid-excreting
habitus.

Once this concentrating kidney had been evolved
as an adaptation to aridity, it supplemented warm-
bloodedness as an adaptation to frigidity, and the mammals were enabled to compete, dry spell for dry spell,

and cold spell for cold spell, with the more sluggish reptiles. Into whatever area the uric acid-excreting reptiles could migrate, the urea-excreting mammals could follow them; and, when winter forced the reptiles to hibernate, the warm-blooded mammals remained active and alert. It is therefore reasonable to suggest that the long delay in mammalian evolution—from the Permian to the Cenozoic, a period of one hundred and fifty million years—was in part attributable to the slow evolution of the warm-blooded state and the concomitant evolution of the mammalian concentrating kidney.

The unique functional feature of the mammalian kidney —its concentrating power—is accompanied by only two notable anatomical changes. The first of these is in the structure of the tubule. The narrow neck, which in the amphibian nephron is interposed between the proximal and distal segments, has lost its cilia and is variably elongated into what is known as the 'thin segment'—so that we can now speak of three segments: 'proximal,' 'thin,' and 'distal.' The tubule as a whole is bent into a hairpin loop, called the loop of Henle: after the proximal segment undergoes extensive convolutions near the glomerulus, it descends in a more or less straight course toward the interior of the kidney; ultimately it turns sharply to return to the glomerulus of origin, and after further convolutions (distal segment) it joins a collecting duct. The thin segment may be confined to the descending limb of Henle's loop, or (as is shown in Figure 10) it may extend around the loop and for some distance up the ascending distal limb.

The second change concerns the disappearance of the renal-portal system. We have emphasized the importance of this ancient venous blood supply in maintaining tubular function in the fishes, Amphibia, and reptiles, at times when the glomerular circulation had to be curtailed. In the mammals, however, the renal-portal system has been completely discarded and the tubules are

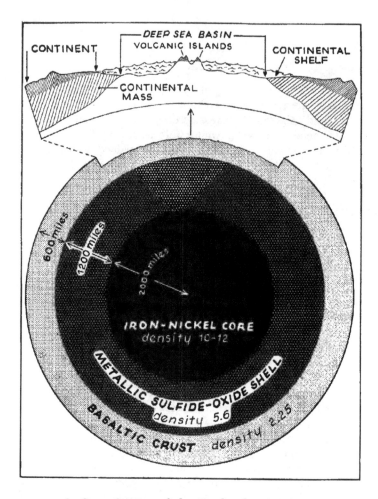

FIGURE 1. Cross Section of the Earth—*showing its concentric shells. The vertical scale of the continental granite is greatly exaggerated; at its greatest depth, this granite does not exceed fifty miles in thickness.*

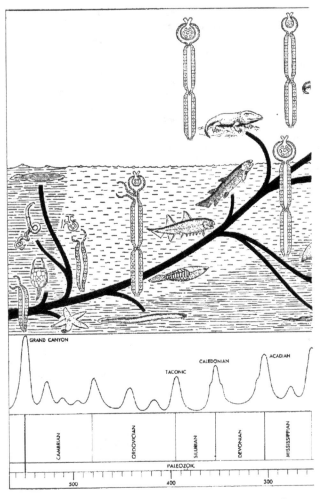

FIGURE 2. Synopsis of the Evolution of Vertebrates—*showing the evolution of the vertebrates in relation to a salt-water (darkly shaded) and fresh-water (lightly shaded) habitat. The irregular curve illustrates mountain-building episodes (geologic revolutions) which have importantly influenced this evolutionary history.*

The time scale is such that the Pleistocene era (one million years in length) and Recent Time (about twenty-five thousand

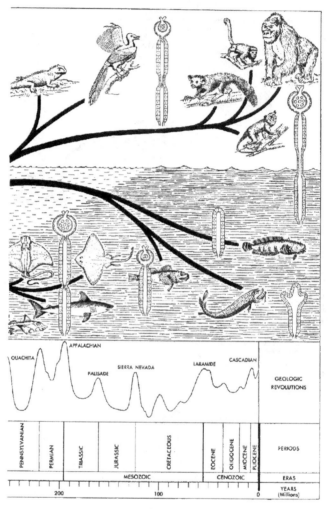

OUACHITA APPALACHIAN SIERRA NEVADA LARAMIDE CASCADIAN GEOLOGIC REVOLUTIONS

PALISADE

PENNSYLVANIAN	PERMIAN	TRIASSIC	JURASSIC	CRETACEOUS	EOCENE	OLIGOCENE	MIOCENE	PLIOCENE	PERIODS
		MESOZOIC			CENOZOIC				ERAS

200   100   0   YEARS (Millions)

years) could not be included, and these are merely suggested by the heavy line at zero time.

The entire period encompassed by documented history is only about six thousand years, or one hundred-thousandth of the interval elapsing since the opening of the Paleozoic era (Cambrian period), when fossilized animals first begin to appear in the sedimentary rocks.

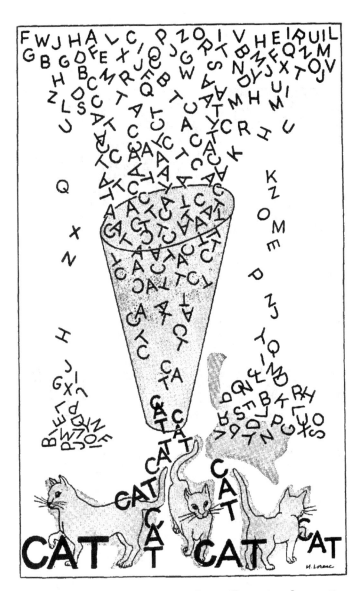

FIGURE 3. Simpson's Alphabet Analogy—*illustrating the creative force of natural selection, but utilizing a funnel instead of a barrel.*

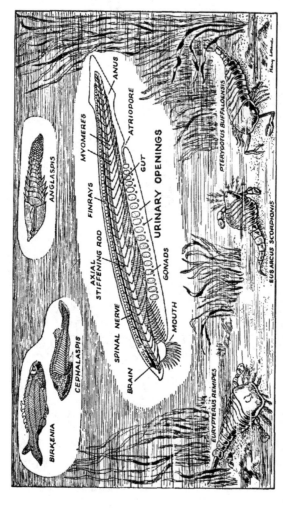

FIGURE 4. (Center): The Hypothetical Protovertebrate (any resemblance to Amphioxus is purely coincidental). (Top): Three Types of Armored Fishes or Ostracoderms from the Devonian. (Bottom): Three Types of Eurypterids from the Silurian. The eurypterids were co-inhabitants of fresh water with the early armored fishes. (Not drawn to scale.)

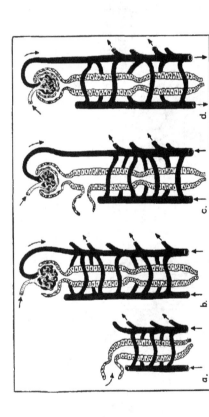

FIGURE 5. Four Stages in the Evolution of the Vertebrate Nephron—(a) In the protovertebrate the renal tubule drained the coelom or body cavity by means of an open mouth or coelomostome. (b) The glomerulus was evolved in the earliest vertebrates as a device to excrete water, and was at first only loosely related to the coelomostome. (c) Later the glomerulus became sealed within the end of the tubule, the coelomostome persisting in some species. (d) In the higher vertebrates, the coelomostome has disappeared entirely, leaving the typical vertebrate nephron. The primitive blood supply to the protovertebrate tubule persists as the "renal-portal system" in the fishes, Amphibia, reptiles and birds (a to c), but disappears in the mammals (d), leaving the tubules supplied only by post-glomerular blood.

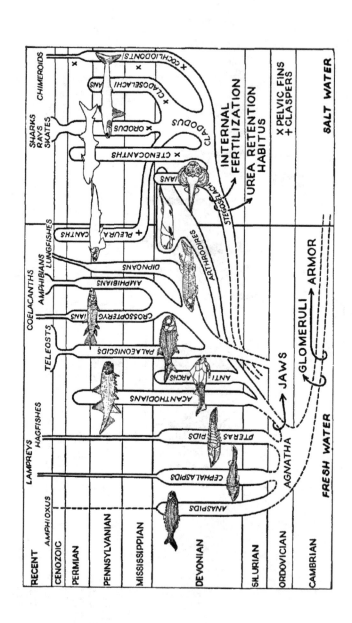

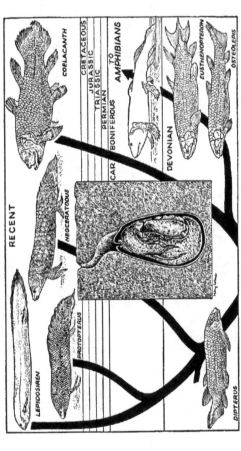

FIGURE 7. The Lungfish, *Protopterus*, in Estivation in a Block of Mud. *The recent lungfishes, Protopterus, Lepido-siren, and Neoceratodus, were evolved from Devonian forms akin to Dipterus, in which the fins had begun to elongate into a rope-like appendage. The Amphibia were evolved from closely related crossopterygian fishes, such as Osteolepis and Eusthenopteron, in which the stubby, paddle-like fin could be rotated under the body and used for locomotion on land. (Not drawn to scale.) The living marine coelacanths, of which many specimens have been caught in recent years, are also closely related to the Devonian crossopterygians.*

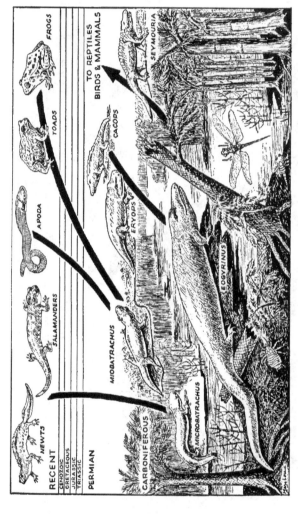

FIGURE 8. The Carboniferous Amphibia—these were evolved from a crossopterygian stem closely related to Eusthenopteron, whose fin was ideally suited for transformation into the tetrapod foot (see Figure 7). (Not drawn to scale.)
The Apoda are blind, burrowing, legless Amphibia found only in the tropics.

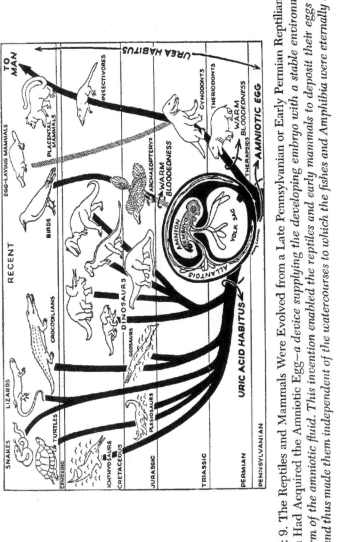

FIGURE 9. The Reptiles and Mammals Were Evolved from a Late Pennsylvanian or Early Permian Reptilian Stock Which Had Acquired the Amniotic Egg—a device supplying the developing embryo with a stable environment in the form of the amniotic fluid. This invention enabled the reptiles and early mammals to deposit their eggs on dry land, and thus made them independent of the watercourses to which the fishes and Amphibia were eternally bound.

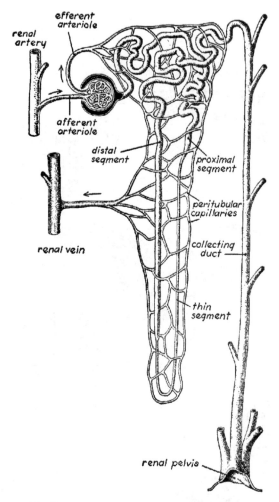

FIGURE 10. Diagrammatic Representation of the Mammalian Nephron. *The blood to this nephron is supplied entirely by the renal artery. After passing through the filtering bed of the glomerular capillaries, this blood is distributed to a rich plexus of capillaries intimately applied to the tubules, from which it is collected and returned to the systemic circulation by way of the renal vein. Each human kidney contains nearly one million such nephrons, all of which are similar in structure and function.*

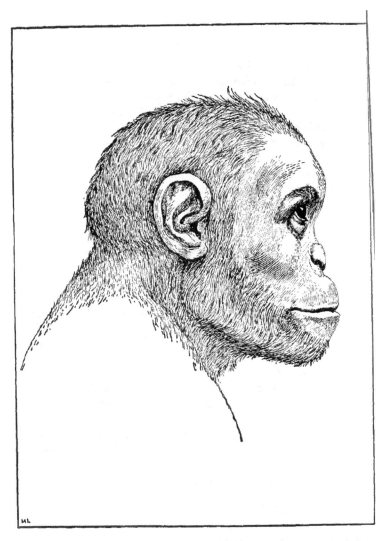

FIGURE 11. Australopithecus africanus — *as reconstructed by G. Elliot Smith and A. Forestier. (Courtesy of the* Illustrated London News.)

supplied with blood only through the glomerular arterioles. After traversing the glomerular capillaries, the blood emerges through a short 'efferent' arteriole which distributes it to the peritubular capillaries, from which it is collected into a meshwork of veins and returned to the systemic circulation. Perhaps the best reason that can be given for the disappearance of the renal-portal system is that, as glomerular function received increased emphasis in the early mammals, and as the filtration rate came to be stabilized at a high level, an independent blood supply to the tubules became superfluous.

When the Cenozoic opened the mammals began to appear in greater variety and abundance, and by the end of the Eocene all the modern orders were represented on the stage, warm-blooded, active, and, for the most part, alert for all the year. These modern mammals are usually divided into 15 orders, as represented by: 1) the shrews and hedgehogs; 2) the armadillos and sloths; 3) the scaly anteaters; 4) the rodents; 5) the picas, hares, and rabbits; 6) the bats; 7) the hyenas, cats, dogs, bears, raccoons, pandas, sea lions, walruses, and seals; 8) the whales, porpoises, and dolphins; 9) the conies; 10) the elephants; 11) the manatee and dugongs; 12) the aardvarks; 13) the horses, tapirs, and rhinoceroses; 14) the pigs, hippopotamuses, camels, deer, antelopes, sheep, and oxen; and 15) the tree shrews, lemurs, tarsiers, monkeys, apes, and man.

The over-all function of the mammalian kidney can readily be visualized in terms of its operations in man, the best-studied species. The two almost identical kidneys in man each contain about one million nephrons. Each nephron consists of a glomerulus and its subjoined tubule, this tubule being differentiated into a proximal segment, thin segment, and distal segment. These tubules drain into confluent collecting ducts that empty into the renal pelvis, from which the urine drains by way of the two ureters into the bladder. (Figure 10.)

The total filtering surface in these two million glomeruli is estimated to be about 0.76 sq. meters, or nearly half the average body surface area (1.73 sq. meters). The filtration rate in man (neglecting sex differences, since all aspects of renal function are slightly greater in males than in females) averages about 125 cc. per minute, or 180 liters (roughly 190 quarts) per day. To supply this enormous quantity of filtrate, the kidneys require nearly 1200 cc. of blood per minute or 1700 liters (1800 quarts) per day. This amounts to one-fifth of the total blood pumped out of the heart, though the two kidneys represent only one-half of one per cent of the total body weight—the kidneys get forty times as much blood per unit weight as does, on the average, the rest of the body.

The 180 liters of glomerular filtrate formed each day contain some 1100 grams (2.5 pounds) of sodium chloride, of which only 5 to 10 grams are excreted in the urine—95 per cent is reabsorbed by the tubules. Some 425 grams (nearly a pound) of sodium bicarbonate and 145 grams of glucose are filtered, and more than 99 per cent of both are reabsorbed. Also filtered, only to be reabsorbed, are substantial quantities of potassium, calcium, magnesium, phosphate, sulfate, amino acids, vitamins, and many other substances valuable to the body.

If the two million tubules in the two human kidneys were connected end to end, they would stretch for nearly fifty miles. There is no reason for surprise, therefore, that by the time this filtrate has emerged from its convolutions into the renal pelvis and bladder it is reduced from 180 liters to an average of 1.5 liters per day of final urine in which waste products—and other substances present in the blood in excess—have been concentrated, some of them a hundredfold or more.

The total extracellular fluid (plasma plus interstitial fluid)—or what Claude Bernard called the 'internal en-

vironment' of the body—in a man of 70 kg. (154 pounds)
amounts to about 11 liters: excluding the blood cells,
proteins, and fats, which do not pass through the glomerular membranes, this entire internal environment is
filtered through the glomeruli only to be reabsorbed by
the renal tubules, some 16 times per day.

Once the glomerular filtrate is in Bowman's capsule
surrounding the glomerular capillaries, it is literally outside the body, because this capsule drains freely into
the renal tubules, the renal tubules into the collecting
ducts, the collecting ducts into the renal pelvis, the pelvis
into the ureters and the bladder. What engineer, wish-
ing to regulate the composition of the internal environment of the body on which the function of every bone,
gland, muscle, and nerve depends, would devise a
scheme that operated by throwing the whole thing out
sixteen times a day—and rely on grabbing from it, as it
fell to earth, only those precious elements which he
wanted to keep? Only nature can be so extravagant, and
only in the light of historical perspective can we understand her extravagance.

It is scarcely necessary to emphasize again that the
substances present in largest amounts in the filtrate, and
reabsorbed in largest amounts by the tubules, are sodium
chloride and water. The evidence from experimental animals and man indicates that some 85 per cent of the
sodium chloride is actively reabsorbed in the proximal
segment of the tubule, and that an equal fraction of
water is simultaneously reabsorbed by passive diffusion,
so that the urine throughout the length of this segment
retains the same osmotic concentration as the blood

One of the most notable advances in renal physiology
of recent years is the demonstration of the role of the
thin segment and the loop of Henle in making an osmotically concentrated urine This is achieved by the
reabsorption of sodium chloride in the ascending portion
of the thin segment in the loop, this sodium chloride,
transferred to the interstitial fluid, increases the osmotic

pressure of this fluid and thereby draws water out of the urine as the latter passes down the collecting duct. Additional sodium chloride is reabsorbed by the distal convoluted tubule and the collecting duct. In the presence of the antidiurectic hormone (ADH) of the pituitary gland all the osmotically free water generated in the tubule by the reabsorption of sodium chloride in the thin segment, distal convoluted tubule and collecting duct escapes by diffusion across the tubular epithelium, so that the urine emerges from the collecting duct with an osmotic pressure equal to that of the interstitial fluid around the loop of Henle. Hence under conditions of dehydration (antidiuresis) the urine flow is minimal (oliguria) and its osmotic concentration maximal.

In the absence of ADH (*i.e.*, during hydration), the permeability of the distal tubule and the collecting duct to water is greatly decreased, and consequently osmotically free water remains unabsorbed and is excreted to form a dilute urine (water diuresis). This theory has two attractive features: it disposes of any 'active' transport of water molecules by the tubule, and qualitatively the same mechanism (sodium chloride reabsorption) is involved in making both a concentrated and a dilute urine. In discussing the absorption of water by the skin in the Amphibia, it was noted that the action of pituitary extracts has been attributed to the dilatation of 'pores' through which water can diffuse: similarly, the dilatation of 'pores' in the distal tubule and collecting duct seems adequate to explain the increased reabsorption of water in the mammalian nephron under the action of ADH.

As one ingests greater or lesser quantities of water, minute changes in the osmotic pressure of the blood work through the osmotically sensitive receptors in the midbrain, the pituitary gland, and ADH secretion, to increase or decrease the excretion of water as required, the system operating so smoothly that the osmotic pressure of the blood generally varies by no more than 1 or 2 per cent. After a single large drink, diuresis starts as soon as the water begins to be absorbed from the intestinal tract, reaches its maximum in 30 minutes, and within an hour or so the body is back in water balance. It is, however, possible to drink water faster than the kidney can excrete it. Because of the division of water reabsorption between the proximal and distal systems, the urine flow cannot exceed the fraction (15 per cent) of the glomerular filtrate that is delivered from the proximal to the distal system: meaning, in an average man with a filtration rate of 125 cc. per minute, roughly 20 cc. per minute or 27 liters (7 gallons) per day. Record beer drinkers consume from 15 to 20 bottles (5.4 to 7.2 liters) in 3 hours (which works out at 30 to 40 cc. per minute) but they end the evening excessively hydrated, that is to say nearly half the water is still in the body when the bar closes.

Alcohol, which exerts an inhibitory action in the central nervous system generally, also inhibits the secretion of ADH, and consequently when taken in concentrated form it increases water excretion, the effect (in physiological doses) being such that 1 cc. of alcohol causes the excretion of approximately 10 cc. of water. Beer containing 4.6 per cent of alcohol is not dehydrating since an excess of water over alcohol is ingested. Straight whiskey, however, with 40 to 45 per cent alcohol, is dehydrating, and consequently it will not serve to quench a man's thirst, for which purpose it must be diluted with water at least one to four.

If water were always freely available, urea and other waste products could readily be excreted either in the dilute urine characteristic of water diuresis or in more concentrated urine formed during the dehydrated or hydropenic (*hydro* = water; *penia* = poverty) state, even though in the latter instance the osmotic concentration of the urine did not exceed that of the blood. But water is not always freely available, nor has it always been

freely available to the mammals during their evolutionary history; and to excrete a urine no more concentrated
than is the blood is uneconomical, in that it entails excessive water loss. Here is where the mammalian capacity for concentrating the urine to an osmotic level
above that of the plasma comes into operation.

The quantitative importance of this concentrating
process can be illustrated by a simple calculation on an
individual who is dehydrated and in whom, therefore,
all antidiuretic mechanisms are operating maximally.
Out of each 100 cc. of filtrate let us say that 97.6 cc. of
water and its contained salt can be reabsorbed without
raising the osmotic concentration of the urine to a level
greater than that of the blood. Such substances as are
not reabsorbed by the tubules will not have been concentrated 100/2.4, or forty-two times, but their excretion involves water loss at a rate of 2.4 cc. per minute, or
3456 cc. per day. In the final process of water reabsorption, however, let us suppose that an additional 1 8 cc.
of water is recovered by the tubules, raising the urine
osmotic concentration to a value 4.0 times (2.4/0.6)
that of the plasma (the maximal osmotic urine/plasma
concentration ratio in man ranges from 3.8 to 4.2). After
this operation the minimal urine flow becomes 0.6 cc.
per minute or 884 cc. per day, a saving of 2572 cc. per
day. Unreabsorbed waste products would now be concentrated 100/0.6 or 167-fold, and, at a urine flow of
884 cc. per day, the water saved by concentrating the
urine would—so long as the kidneys alone are considered
—prolong life in a man wholly deprived of water by
three days. Yet even when water is freely available, the
urine flow in man averages only about 1500 cc. per day,
so that his kidneys are generally operating on the
moderately 'concentrated' side.

CHAPTER XI

ANIMALS THAT LIVE WITHOUT WATER

In water-conserving ability, man compares poorly with many other animals, some of which can subsist with no source of water other than the moisture of their food (preformed water) and that generated by the oxidation of this food in the body—so-called 'metabolic water.' We have briefly mentioned metabolic water in connection with the lungfish, reptiles and birds, but did not enlarge on the topic because of lack of information on its importance to these forms. In the present connection, however, the term warrants detailed definition. Each of the three foodstuffs, carbohydrate, fat, and protein, contains a high proportion of hydrogen to carbon; as oxidation converts the carbon to carbon dioxide, the hydrogen is oxidized to water: one hundred grams of dry starch on oxidation yield 55.6 cc., one hundred grams of fat yield on the average 107 cc., one hundred grams of protein, if the nitrogen is degraded to ammonia, urea, or uric acid, yield, respectively, 32, 39.6 and 53 cc. of water. On a balanced diet (3000 calories) of 465 grams of carbohydrate, 85 grams of fat and 90 grams of protein, the metabolic water in man would amount to 335 cc. per day. But under the most favorable conditions man loses 700 cc. of water per day through his skin and lungs; hence, taking the minimal urine flow on a mixed diet as 900 cc., his minimal additional water requirement is roughly 1300 cc. per day, and consequently he must supplement his dry food with liquid.

Some animals, however, are so well specialized for arid life that they can live on the metabolic water of their food plus such preformed water as is invariably present if the food is not oven-dried. One of the first students to see the importance of this problem was S. M. Babcock of the Agricultural Experimental Station of the University of Wisconsin. Babcock was at the time concerned with the viability and germination of seeds, but secondarily he was led to demonstrate that clothes moths, grain weevils, dry-wood borers, bee moths, and others, as well as their larvae, subsist wholly upon metabolic water. He showed that clothes moths will live and lay viable eggs when kept in a desiccator the air of which has been dried over sulfuric acid, and when feeding on a piece of oven-dried woolen cloth; and that they will live when fed on dry mink or astrakhan—the second generation dying of starvation only when every particle of fur has been consumed, leaving the white, clean skin. The larvae at various stages contained from 57.7 to 59.8 per cent water, the woolen cloth or fur only 6.1 to 9.1 per cent—in other words the animal literally manufactured water from its food. The larvae of the bee moth (containing from 57.3 to 59.2 per cent water) can live on the dry wax of the honeycomb, which contains less than 2 per cent water, while obtaining its nitrogen from adhering pollen grains. And so it is with the pea weevil, the confused flour beetle, the flour moth, the tobacco hornworm—all live on food containing less than 10 per cent preformed water.

A well-dried piece of mink, astrakhan, or wool as the sole source of water may well be taken to epitomize 'desert life,' but the reptiles and mammals have invaded desert areas in a broader and more active sense. Biologists use the word 'desert' to designate places in which the climate is hostile to animals and plants (though few are the desert areas that are actually uninhabited by both), in which normal agriculture is impossible, and in which nearly all indigenous animals and plants are specialized to endure continuous aridity. It is estimated that by this definition as much as 20 per cent of the earth's surface may be called desert. With slight exceptions along the coasts and river valleys, a desert climate prevails over the whole of northern Africa and eastward to northwestern India and the heart of China, reaching maximal aridity in Egypt, Saudi Arabia, Syria, Mesopotamia, Persia, Afghanistan, Baluchistan, Turkistan, and the Gobi Desert (collectively called the Great Palearctic deserts). The desert of Australia is second in point of size, and deserts occupy large areas in the western part of North America and in Mexico, eastern Patagonia, western Argentina, and southwestern Africa.

The desert is characterized by extreme aridity and sometimes by high mean temperatures, and for a good part of the year the prevailing winds may be dry and desiccating. A rainfall of less than 5 inches per year is typical, but a greater rainfall may be offset by persistent hot, dry winds. The rainfall is invariably distributed unevenly throughout the year, the rainy season varying with the latitude, prevailing winds, and the disposition of the neighboring mountains. In some areas rain falls on only a few days in the year and is so variable that a single day's downpour may exceed the total rainfall of an unusually dry year. Because the ground lacks cover, a great proportion of this water runs off immediately and is of little use to living organisms. But all soil—whether gravel, sand, sun-baked mud, loam, or peat—has a remarkable capacity to hold water and to raise it from lower levels by capillary action, and consequently the subsurface water is generally a more important factor in sustaining plants than is the rare rainfall.

In the absence of clouds and moisture, the daytime temperatures in the summer may rise well above 125° F. in the shade, while at night the temperature falls rapidly and even in summer may approach the freezing point. Only the superficial earth, however, suffers these extremes, and at a depth of 6 inches the daily fluctuation in temperature is reduced to a quarter of that of the air; at 10 inches, to a thirtieth. Because the desert temperature falls rapidly, the dew point is frequently reached within an hour or so after sunset, and the dew that accumulates on plants and grasses is sometimes a significant source of water for many animals. There is, however, considerable variation in the humidity of desert areas—the formation of dew in the Arizona desert, which is protected by mountains, is a rare event, whereas the annual precipitation of dew in the Negev of southern Israel, which receives moisture-laden winds from the Red Sea and the Mediterranean, amounts in the average to 8.7 inches, or more than many deserts receive as rain.

Desert plants fall into three general groups. The annuals sprout quickly in the rainy season, bloom in an incredibly short time and then wither rapidly after reseeding the soil. It is these ephemeral annuals that give to the desert its many colored flowers for a short period of the year. The second group is represented by perennials that have a bulb, tube, or fleshy root buried fairly deeply in the soil; these sprout and multiply during the rainy season, though less rapidly than do the annuals. Neither of these two groups store significant quantities of water in the fleshy parts. The third and most important group, from the year-around point of view, is represented by the succulent plants such as the cacti of the Western Hemisphere or the spurges (*Euphorbia*) of the Eastern Hemisphere. In both, the stems are greatly enlarged and leaves are small or absent; stems and leaves are tough and leathery and the thickened cuticle may be covered with wax, resin, or densely matted hairs, and in all cases the enlarged stems act as reservoirs for the storage of water. A large tree cactus might contain as much as a thousand gallons, and a donkey can be watered with the juices of the giant cactus, *Cereus gigan-*

teus (which is too bitter for man), while the juice of the barrel cactus (*Echinocactus*) may furnish several pints of fluid which even man can drink. Many if not most of the succulent plants are thorny, and though it may be thought that this thorniness is a protective adaptation against thirsty animals, it is more likely that it is merely an outgrowth of the horny, hypertrophic cuticle and that any protective value is incidental.

As the annual and tuberous-rooted plants show marked seasonal activity, so also do a great variety of animals. Between the rains, the grasshoppers, beetles, and other insects disappear into the earth, and in many species the reproductive cycle places them in an inactive larval or pupal stage below the ground in the dry season, to appear again, sometimes in abundance, with the short-lived annual vegetation. Some are abundant in midsummer, not because midsummer conditions suit them particularly but because several months before there had been food and water for the larvae. Ants, which are generally insect eaters, tend to follow the rest of the insect cycle, but many live on plant juices or the secretions of plant lice which are accumulated during a period of abundance. Specialized workers store these foods in the stomach, and become so distended that they are known as honey-pot ants.

Lizards and other desert reptiles can probably subsist on the preformed and metabolic water of the insects which constitute their food, but no reptile, so far as is known, can live on dry food such as seeds. The Amphibia are at an even greater disadvantage: frogs and toads must burrow into the earth where there is substantial humidity, emerging only at night, and during the dry season they pass into a state of relative inactivity below the ground, sleeping in the subsoil moisture.

The migratory birds, living on ants, insects, frogs, or lizards, may invade the desert in increasing numbers during the lush season, but relatively few have established themselves as permanent residents. The sand grouse is reported to hatch its eggs when the air is very dry and the desert soil too hot to touch, the vegetation parched, and the seeds all but desiccated. The adults fly long distances to watering grounds at regular intervals, and to sustain the nestlings, the male, before drinking, rubs his breast violently up and down on the ground to ruffle the feathers and then wades in to saturate his underparts; when he returns to the nest the young, which live on dry seeds supplied by the parents, get under him and suck the water from the feathers by passing them through their bills. Even more than the sand grouse, the thrushes and desert larks appear to have achieved almost complete independence of a supply of drinking water. The ostrich is perhaps as independent as any bird and roams many miles away from water. Whereas other birds, whose eggs are smaller and heat more rapidly, must nest upon them during the day to keep them from being cooked, the ostrich buries its eggs in exposed sand, the surface temperature of which may reach 125° F. or higher—presumably only the large size of the egg keeps it from being roasted.

The desert invariably presents an elaborately interrelated pyramid of vegetable and animal life. At its worst, when moisture is at the greatest premium, it may seem to be devoid of any living thing except cacti, dried-up sagebrush and a few scrubby plants; but even then one can generally discover under a large flat stone a population of wood lice, centipedes, millepedes, spiders, scorpions, mites, cockroaches, crickets, beetles, bugs, ants, snails, and earthworms, and even an occasional lizard, snake, or frog. Then, with the first increase in moisture, these creatures venture from under cover and the woodpecker comes to feed upon them, and to hew out a nesting place in the giant cactus; and when the woodpecker abandons the nest after its breeding season, owls and other birds come to occupy it, honeybees use the empty nests as hives, insect scavengers survive in both bees' and birds' nests, until at the peak of the season the once seemingly barren region is well populated. Each new inhabitant brings in its wake those who prey upon it, until even coyotes, foxes and other carnivores can invade the waterless wastes with some degree of safety.

A favorable year climatically is accompanied by enhanced growth and numbers of all animals; an unfavorable year reduces both the adult size and the number of offspring and, if drought is abnormally prolonged, the species may be so reduced that several years are required to re-establish it. There is no evidence that the fertility or reproductive rate in any species whatsoever is stimulated by adverse conditions.

It is not, however, among the carnivores that we find the most successful mammalian invaders of the desert, but among the herbivores which live directly on the desert vegetation. The reason for this seems clear enough: except for those that hibernate in cold weather, the mammals, unlike the insects, must be active the year around; and although some of them may migrate to more favorable areas, as do the birds, they must in general reproduce and rear their young under strictly desert conditions. To colonize the desert successfully, the mammals must feed directly on the land and not on its seasonal population.

Thus it is that many large, herbivorous mammals, such as the giraffe, eland, antelope, and gazelle, seem able to sustain themselves in desert areas far from any source of drinking water. The gazelle's ability to do without water probably accounts for the fact that it is seen more frequently than any other mammal in many of the world's deserts.

Unfortunately, there are almost no rigidly controlled observations in these matters, and much that passes for fact is mere legend. The legend of the camel, for example, is notorious, but the legend is unquestionably exaggerated. There is no evidence, and it is highly improbable, that the camel 'stores' water in either its stomach or its hump. The stomach in all ruminants is a complex series of chambers in which food is stored before regurgitation for chewing of the cud, and in which bacterial decomposition can aid in the digestion of hard grasses. Misinterpretation of the presence of digestive juices in these chambers is responsible for the stomach water-storage legend. The camel's hump is, however, a large reservoir of solid fat which, by combustion, can yield twenty to thirty pounds of water. The water required for urine is reduced by subsistence on carbohydrate-rich, protein-poor food, since there is less nitrogen demanding water for its excretion; and if food is scarce, the animal's water requirement will increase because of the deficiency of metabolic water, while working will, of course, further cut down its endurance. In midsummer, and with even moderate marching, eight days without water is probably close to the camel's limit. It has been said that the camel can drink 'salt water,' which man and horses cannot drink, but this does not imply that they can drink undiluted sea water—travelers frequently call brackish water 'salt' even though the salt content is fairly low from the point of view of over-all water economy, and such reports must be taken cautiously. The camel is said to eat plants no matter how spiny, and to prefer some of the more thorny shrubs to others which are unarmed but which may be bitter or nauseating—a tale more credible than most of the camel legends. One notable adaptation of the camel is that when exposed to the hot sun the body temperature increases by as much as 12° F., thus substantially reducing the water lost through the skin and respiratory tract.

In warm weather cattle require water, but sheep can remain in pasture without drinking for weeks—though watering improves their growth; and in the great Australian desert, where much of the land is devoted to sheep raising, the animals must be watered from artesian wells. The wild ass is believed to subsist without drinking water, but the domestic donkey cannot; after an arduous day's labor the donkey compensates for water loss by taking a fabulously large drink, consuming as much as 12 per cent of its body weight in the space of five minutes.

In mammals that are physically active in the hot sun, the control of body temperature by the evaporation of water—whether as sweat, as in man, horses, and cattle; by panting, as in the dog; or both, as in the sheep—may increase the total water requirement far above that involved in urine formation. Sweating presents the additional complication in that it may also cause the loss from the body of considerable quantities of salt. In animals that sweat profusely this loss of salt is such as almost to keep pace with water loss, and the blood does not become unduly concentrated. Consequently, man, for example, quenches his thirst with a little water and only restores the total volume of his body fluid over a period of hours or days as the lost salt is replenished through his dietary intake. If he drinks too much water all at once when in the dehydrated state he may in fact suffer aches and pains from rapid overdilution of the blood. The reason the donkey satisfies its thirst in one big drink (and the dog is rather more like the donkey than like man) is that it does not sweat to so great an extent, and consequently loses little salt, and the loss of water in respiration leads to concentration of its blood; consequently it drinks enough at one time to bring itself back into water balance.

Most of the mammals which are known authentically to live in the wild state without access to drinking water are of small size. The ground squirrel and pack rat live on succulent vegetation and cacti containing from 60 to 90 per cent water, the grasshopper mouse on insects containing 60 to 85 per cent water. The common wild hares of the United States, *Lepus alleni* and *L. californicus* (incorrectly called 'jack rabbits'), and the prairie dog also live on succulent plants; they can be maintained in captivity on a mixture of greens and dry foods, but not on dry food alone. All of them will drink water when available, especially if they are kept on a dry diet.

At the extreme, the mammals most successful in living without free water are the kangaroo rats: *Dipus, Jaculus, Gerbillus, Meriones, Dipodillus* of the Great Palearctic deserts, *Pedetes* of South Africa, *Dipodomys* and *Perognathus* of the American deserts, and *Notomys* and *Ascopharynx* of Australia. Though belonging to different families, these rodents are all small, they all have elongated hind legs and a long tail, and they jump in a kangaroolike manner, from which habit the group derives its popular name. Several of them have cheek or throat pouches for carrying seeds, in the manner of the squirrel, and they are generally as skillful as the squirrel in using the front paws as hands.

Several of the American kangaroo rats, *Dipodomys merriami* and *D. spectabilis,* and Bailey's pocket mouse, *Perognathus baileyi,* have in recent years been extensively studied by Knut and Bodil Schmidt-Nielsen and are the only desert-living mammals on which detailed physiological information is available. These small animals normally live on air-dried seeds and can apparently grow and reproduce and nurse their young on a diet of dry grain or oatmeal containing only 5 to 10 per cent of free water. Actually, they have to be taught to drink water, and while *Dipodomys* will eat succulent food, *P. baileyi* disdains green things and will not even eat fresh watermelon.

Dipodomys has no mechanism for storing water in the body when it is living on moist food, nor does it lose significant quantities of water during a long period on dry food. It can, however, be forcibly dehydrated by feeding it a high-protein diet of dried soybeans (40 per cent protein), which causes increased loss of water in the urine in consequence of the large quantity of urea requiring excretion. When dehydrated in this manner, it dies in two or three weeks with an average weight loss of about 34 per cent, but with no significant decrease in the actual water content of the body: death is a result of a decrease in the volume of body fluid rather than a change in composition.

Whereas, in some uric-acid-excreting insects, water may be saved from excretion by storing uric acid in the body throughout the life of the individual, in the kangaroo rat (as in all other mammals), in which protein nitrogen is converted to urea, no such storage mechanism exists and the urea is excreted as fast as it is formed. So long as the filtration rate remains at a constant value, this urea, with sodium chloride and other urinary constituents, automatically sets the minimal rate of urinary water loss. Hence the degree to which the urine can be concentrated becomes of paramount importance. The kidney of the kangaroo rat can concentrate to the greatest extent of any known mammal, reaching osmotic concentrations 17 times that of the plasma, as compared with the maximal figures of 5.6 in the seal, 4.5 in the dog, and 4.2 in man. The concentration of urea in the urine reaches 23 per cent, that of sodium chloride 8 7 per cent, as compared with about 6.0 and 2.3 per cent, respectively, in man. The urine is, in fact, so concentrated that it is apt to solidify after it is withdrawn from the bladder.

This extraordinary concentrating power enables *Dipodomys* to drink even sea water, the only mammal on record that can do this. But to induce it to drink sea water, which it finds unpalatable, it is necessary to dehydrate it first by feeding it a diet so high in protein nitrogen that it cannot otherwise maintain itself in water balance. Since the kangaroo rat can concentrate sea water two and a half times, something more than one-half of the water of the sea water is in theory available for urine formation. It would be expected that the poorly absorbable salts, magnesium and sulfate, would cause diarrhea, but diarrhea is not reported, which is the more surprising since the animal drank quantities of sea water ranging from 3 to 15 per cent of the body weight per day.

If an aglomerular mammal were possible one would expect to find it here, but the glomeruli of the kangaroo rat are typically mammalian, and the rate of glomerular filtration is of the same magnitude per unit of body weight as in the laboratory white rat, an animal not particularly specialized for arid life. Calculated on the basis of oxygen consumption, the filtration rate in the kangaroo rat ranges from 0.27 to 0.48 cc. per cc. of oxygen, figures to be compared with the range of 0.35 to 0.75 cc. in man. Clearly the filtration process is not severely curtailed, despite the fact that the average normal urine output is only one-tenth (0.0004 cc.) of that in man (0.004 cc. per cc. of oxygen consumed).

Perhaps the rodents are geologically too young to permit degeneration of the glomeruli, but an alternative explanation is more attractive: when, in the evolution of the mammalian kidney, the renal-portal system was abandoned, the kidney got into a blind alley of its own —the filtration-reabsorption system is now so firmly established that there is no easy way to overhaul it and to convert it to a purely tubular kidney, as the marine fishes have done. The crisis of water deprivation can be met only by making a more and more concentrated urine and conserving water to the utmost by every other means —which is just what the kangaroo rat does, breaking all records in these respects.

Calculated on the basis of oxygen consumption, man loses much more water by evaporation from the lungs alone than *Dipodomys* loses from the lungs and skin together. One of the adaptations reducing respiratory water loss in *Dipodomys* is a long nose, because the longer the nose, the lower the temperature at the end and the less water lost by evaporation into the expired air. *Dipodomys* has no longer a nose than the white rat, but the latter suffers the disadvantage of greater water loss through the skin despite the fact that, like *Dipodomys*, it does not sweat. A long nose and dry skin contrive to keep total extra-renal water loss to only slightly more than half that of the white rat, and to probably one-third or less than that of man under moderate, resting conditions.

When total water loss (that which is evaporated from the lungs and skin and excreted in the urine, plus a small loss in the feces) is balanced against total water intake (the preformed water of the food plus metabolic water), the issue of net water balance—the issue of life or death —hinges importantly on the relative humidity of the atmosphere in which the kangaroo rat lives. *Dipodomys* has excess water available to it at all relative humidities above 10 per cent. During the driest part of the year the humidity above the ground in the Arizona desert is usually near this critical limit, and consequently the animal cannot afford to risk loss of water by exposure to the direct sun: with air temperatures as high as 104° F. and ground surface temperatures reaching 157° F., it must remain in its burrow except at night. The average depth of the burrow is only two feet, but the temperature rarely exceeds 86° F. and the humidity averages 30 per cent. The humidity in a small, closed pocket of air at a depth of only 12 inches is 100 per cent, but the humidity of the burrow is reduced by ventilation, and some species keep ventilation to a minimum by closing the entrance of the burrow during the daytime hours.

The humidity of the burrow is no less important in respect to the storage of food. Pearled barley, with which these quantitative observations were carried out, yields 53.6 cc. of metabolic water per 100 grams. If dried in the desert air, the free or preformed water would be negligible, but after storage at a humidity of 30 per cent the free water increases to 10 cc. per 100 grams, increasing the total available water to 63 cc. This increase in available water is significant, because no potential source of water gain can be ignored. Nor can any potential source of water loss be ignored—to which end, in some species of kangaroo rats, the animal's cheek pouches, by which it carries home its sun-dried food, are lined with fur.

Man's water requirements in the desert assumed potential military significance in World War II, and were studied extensively by a group from the University of Rochester under the direction of E. F. Adolph. By far the greater part of the water lost by man at high temperatures is in the form of sweat, and sweating is, of course, highly variable, depending on temperature, humidity, and physical activity. At 100° F. water loss by sweating (including respiratory loss) ranges from 7.2 liters per day, when the subject is sitting clothed in good shade, to more than 24 liters per day when he is walking clothed in the sun and carrying a 33-pound pack, and to 28 liters per day when walking naked in the sun without a pack. (These figures are to be compared with one liter or less as the minimal quantity of water required for urine excretion.) Since sweat contains considerable quantities of salt, men sweating under desert conditions eat excessive quantities in order to compensate for salt loss, the daily loss reaching, in the extreme, 10 to 15 per cent of the total quantity in the body. Some men accumulate a visible crust of salt on the skin during a day's activity in the sun. One of the notable adaptations to a hot climate (and to hot weather) is that the salt content of the sweat diminishes after a week or so of sweating, though never enough to prevent significant loss of salt from the body. But even under extreme conditions, the kidneys maintain the salt:water ratio (and hence the osmotic concentration of the body fluids) within very narrow limits.

Depending on the dietary intake of salt and protein, 700 to 900 cc. per day of urinary water are needed for the excretion of waste products. The figure of 300 cc. per day may be taken as the extreme minimal urine output even on a protein-free diet. In soldiers engaged in routine military activities but allowed all the water they wanted to drink, urine formation averaged only 935 cc. per day—a figure to be compared with 5900 cc. (6.2 quarts) of total water ingested by these same individuals. It seems that even when men can drink all the water they want, if sweating excessively they remain in a slightly dehydrated state, presumably because a continuing small water deficit in the body is necessary to stimulate the drinking of the large quantities of water that are necessary to compensate for the great loss in sweat. In short, it is something of an effort to drink water all the time.

The excretion of urine can be further reduced only as the quantity of urinary solutes is reduced by diet, but dieting, in turn, has very limited possibilities. Restriction of salt intake is inadvisable because, even after adaptation to heat, considerable salt is still lost by sweat, and salt restriction can lead to salt deficiency with serious clinical disturbances. Protein intake can be moderately restricted, but if there is too little protein in the diet nutrition will be impaired. In any case, the fraction of water lost in the urine (700 to 900 cc.) as compared with sweat (7000 cc. and upwards) is so small that such measures are of no practical advantage.

It is believed that men who suffer near-lethal dehydration cease to form urine, probably as a result of circulatory failure and consequent reduction in blood pressure and hence in the filtration rate. There is as yet no evidence that the filtration rate in man can be functionally reduced in order to conserve either salt or water, short of the inevitable reduction in renal blood flow that follows when circulatory failure leads to vasoconstriction throughout the body in order to preserve the blood supply to the heart and brain. Man in particular, and probably the majority of mammals, thus differ from the

Amphibia and fishes in which a sensitive functional adjustment, upward or downward, of the filtration rate is an intrinsic part of the mechanism of maintaining salt and water balance.

Under the field conditions studied by Adolph and his collaborators, complete cessation of urine formation was not observed; and they found no evidence that renal function is impaired by repeated but tolerable exposure to heat or moderate dehydration, or both. They estimate that, at maximal daily shade temperature of 120°, 110°, 90°, 80°, and 70° F., a man with no water available can survive 2, 3, 5, 7, 9, and 10 days, respectively, if he is engaging in no exercise at all. If he is walking at night until exhausted and resting in the shade thereafter, expected survival is reduced to 1, 2, 3, 5, 7, and 7.5 days. These figures are, of course, estimates only, because there are no reliable data on the actual circumstances of death of men under these conditions.

At the other extreme of water deprivation are those mammals which have made the sea their permanent habitat: the cosmopolitan seals, the sea lions of the Pacific, and the walruses, all of which are carnivores and distantly related to the otters; and the whales, porpoises, and dolphins, the ancestry of which is obscure. The herbivorous sea cow, manatee, and dugong, of Ungulate origin, are not strictly marine but live along the coasts and in the rivers.

The seals and whales differ from each other in the fact that the seals come ashore to breed and bear their young, and also to fish and play along the coasts, whereas the whales never come ashore and bear and nurse their young at sea. Little information is available on these marine mammals, but isolated observations on the urine of the rorqual whale (*Balaenoptera borealis*), the blue whale (*B. musculus*), the sperm whale (*Physeter catodon*), the humpback whale (*Megaptera boöps*), the porpoise (*Phocaena sp.*), and the seal

(*Phoca barbata* and *P. foetida*), give no evidence that these animals have superior renal concentrating power or that they can drink sea water, while calculations based on the composition of the diet indicate that they can live on the free and metabolic water of their food. The seals, porpoises, and dolphins eat mainly fish, the killer whale eats the flesh of other whales, birds, and seals; the bottle-nosed whale and sperm whale eat cuttlefish and squid. The whalebone whale (including *Balaenoptera*) lives on plankton (small animals) and fish which it collects by straining sea water through its mouth where the plankton is caught on the blades of whalebone hanging from the palate. There may be as many as three hundred or more of these blades on each side, and those in the middle may reach a length of 10 to 12 feet. At the inner edge they fray out into long, delicate, but tough hairs, and their arrangement is such that they fold back when the mouth is shut, but unfold and completely fill the cavity when the mouth is open, so that all the animal has to do is to open and close the mouth in order to filter out from one to 50 gallons of small organisms. When the mouth is closed the tongue is forced against the palate, expelling most of the water and concentrating the filtered food into a semisolid mass which apparently is well compressed before swallowing, since the urine gives no evidence that the animal takes in much sea water.

The harbor seal, *Phoca vitulina*, is the only one of the marine mammals that has been studied in detail. This is a strictly marine form that feeds on fish that it can swallow either above or below the water; in the latter instance the esophagus wipes the fish virtually dry as it goes down. The seal opens its mouth frequently under water, and sometimes plays with a fish for a long time before swallowing it—mouthing it, letting it loose and mouthing it again—but at no time does it swallow any significant quantity of sea water.

Studies on the seal through the cycle of fasting and feeding show that urine formation is intimately related to feeding. As digestion proceeds and the requirement for urea excretion increases after a meal, and as metabolic water becomes available, the urine flow increases to large values, only to decrease again to 10 per cent or less of the maximal value 8 to 10 hours after the meal, when the available metabolic water has been spent. The composition of the urine is in no way remarkable as compared with other mammals. The maximal osmotic concentration of the urine at minimal urine flow is about 5.6 times that of the plasma, and thus significantly greater than in the dog (4.5) but very much less than in the kangaroo rat (17 or higher).

The increase in filtration rate in the fed animal is not attributable to the opening up of inactive glomeruli, but to an increase in function in all glomeruli. Thus the seal maintains itself in water balance not by excessive concentration of the urine, as does the kangaroo rat, but by reducing the quantity of solutes filtered through the glomeruli. The cycle in urine flow following a meal parallels, and is largely caused by, change in the rate of glomerular filtration, a phenomenon not observed in man. It is probable that this unique control of the renal circulation is related to the fact that the seal is a diving mammal. When it submerges to catch fish it must sometimes engage in vigorous swimming with no possibility of respiration, and L. Irving and his collaborators have shown that the seal and other diving mammals have developed an elaborate reflex that involves arrest of respiration, slowing of the heart, and constriction of the blood vessels in large areas of the body, thus reducing the blood flow to all organs other than the heart and brain and temporarily conserving oxygen; this oxygen debt is paid off by increased respiration when the animal returns to the surface. This diving reflex can be elicited by merely holding the nostrils closed when the animal is out of water. During this artificial dive the renal blood flow and filtration rate (and of course the urine flow) are reduced to very low levels. A considerable quantity of blood is thus diverted from the kidneys and made available for circulation elsewhere. It may be inferred that some sort of a diving reflex is present in all the truly marine mammals, since they all can remain below the water for some period of time, and it is possible that the evolution of the diving reflex paved the way for the evolution of the metabolic control of renal function, the conjoint adaptation making it possible for these mammals to live exclusively in the sea.

So far as is known, terrestrial birds are either drinkers of fresh water or, among desert forms, able to live on water derived from food. The maximal osmotic concentration of the urine in the chicken (unfortunately the only bird studied to date) is only some 60 per cent greater than that of the blood (recall that the osmotic urine/plasma concentration ratio in the kangaroo rat may reach the spectacular value of 17). If this limitation applies to other birds, as seems to be the case from fragmentary data on the maximal concentrations of sodium and chloride in the urine, the availability of water or very moist food must be important in determining their habitat, despite their excretion of uric acid, because salts and nitrogenous substances that cannot be converted to uric acid will demand considerable water for their excretion.

Some birds, however, have solved the problem of spending weeks or months—even years—at sea, returning to the land only to lay and hatch their eggs. Among the truly marine birds are the shearwater (*Puffinus*), the petrel (*Fulmarius*), and the albatross (*Diomedea*); while many coastal birds, such as the gulls (*Larus*), the guillemots (*Uria*), auks (*Plautus*), and cormorants (*Phalacrocorax*) appear to be independent of any freshwater supply. Until recently it was thought that these marine forms were also entirely dependent on metabolic water, and indeed that they would, if they drank sea water, suffer all the adverse consequences of doing so— excessive salt-loading of the body and dehydration in consequence of both the osmotic diuresis caused by the sodium chloride in the urine and osmotic diarrhea caused by magnesium and sulfate unabsorbed from the intestinal tract. Apart from the hazard of diarrhea, the bird kidney can excrete salt in a concentration only about one-half that of sea water, so that drinking sea water would entail the excretion of at least twice as much urine as the volume of sea water ingested.

Then in 1956 Knut Schmidt-Nielsen and his collaborators made one of the most interesting discoveries of modern biology when they undertook the study of the salt and water balance of the double-crested cormorant, *Phalacrocorax auritus*, at Salisbury Cove. On a diet of fresh fish, of which the birds consumed up to half of their weight per day, the derived water proved to be far more than adequate to cover the excretion of salts and uric acid, and no evidence was obtained that sea water was ever ingested, though this was the only water available except in the food. To determine what would happen if sea water were ingested, quantities amounting to about 6 per cent of the body weight were administered by stomach tube. As was to be expected, the concentration and rate of excretion of urine were quickly increased, chiefly in relation to increased excretion of sodium chloride. But what came as a complete surprise was the secretion of a clear, waterlike liquid by two glands in the head which drain into the internal nares and are known to anatomists as 'nasal glands.' This liquid ran from the nasal openings and down the beak to accumulate at the tip from which the drops were shaken off by sudden jerks of the head. The secretion proved to be an almost pure solution of sodium chloride (averaging, in the case of the cormorant, 529 milliequivalents per liter, with some 12 milliequivalents of potassium chloride), the sodium chloride concentration being greater than that in sea water (490 milliequivalents per liter) and having about twice the maximal osmotic concentration of the urine. (The magnesium, calcium, and sulfate of the sea water are apparently poorly absorbed from the intestinal tract, and such of these divalent salts as are absorbed are presumably excreted by the kidneys). The maximal rate of nasal secretion (ca. 0.2 cc. per minute in a 1.5 kg. bird) was such that if continuously maintained it would, together with the maximal excretion of sodium chloride in the urine, suffice to remove all the sodium in the body in roughly 10 hours.

Paired nasal glands are present in both terrestrial and marine birds and have been known to anatomists for many years. In the marine birds, however, each gland is greatly enlarged and possesses a rich arterial supply and a more highly developed glandular structure than in terrestrial forms. Invariably draining through the internal nares, the gland may be located either in the supraorbital groove of the frontal bone or in the orbital cavity close to the interorbital septum. Hitherto the function of the nasal gland has been incorrectly interpreted as protecting the nasal mucosa from sea water by rinsing away the latter when it penetrates into the nasal cavities.

In the herring gull (*Larus argentatus*), in which the anatomical structure is best known, the gland consists of branched, secretory tubules radiating from a central canal or duct; the tubules are made up of cylindrical or polygonal cells characterized by striations or lamellae extending lengthwise, apparently from one end of the cell to the other. The blood supply is mainly from the internal ophthalmic artery, and within the gland highly branched capillaries with frequent anastomoses run radially from the central canal toward the surface, more or less parallel with the tubules. The capillary blood flows in a direction opposite to the secreted fluid in the lumina of the tubules, but whether this 'counter-current' flow is physiologically important is not known. The nerve supply is from a parasympathetic ganglion in the an-

terior part of the orbit, this ganglion communicating
primarily with the VII cranial (facial) nerve. The post-
ganglionic fibers to the gland are probably cholinergic,
and secretion is stimulated by Mecholyl and acetylcholine, and inhibited by atropine. Secretion is also blocked
during anesthesia.

The gland excretes only sodium chloride in hypertonic
solution (with traces of potassium chloride), and it cannot excrete osmotically free water (*i.e.*, a hypotonic solution such as is secreted by the tear glands, etc.). Its
structure is such as to exclude any filtration-reabsorption process, and to demand a purely secretory operation, as in the aglomerular kidney. Unlike the latter, the
salt gland, as Schmidt-Nielsen has called it, secretes a
solution (specifically of sodium chloride) hypertonic to
the blood, and no other organ in the vertebrates is known
to accomplish this except the bird and mammalian kidney (and here hypertonicity is achieved indirectly by the
active reabsorption [secretion] of sodium chloride into
the interstitium of the renal medulla).

The kidneys of the cormorant respond to a load of
pure water by the increased excretion of dilute urine
(water diuresis), and to a load of sodium chloride or sea
water by the increased excretion of osmotically concentrated urine with a high sodium chloride content (osmotic diuresis). However, the capacity to concentrate
the urine osmotically is at best slight (as indicated by
the maximal salt content); consequently the salt gland
serves as a safety device insuring that any excess sodium
chloride, ingested as sea water or in the diet, can be
disposed of without excessive loss of water. The stimulus
which excites secretory activity is an increase in the osmotic pressure of the blood, not in the sodium concentration, because the intravenous administration of hypertonic sucrose solution is as effective in inducing secretion
as is the injection of hypertonic sodium chloride or the
oral administration of sea water. The high osmotic concentration of the secretion permits the salt to be excreted

in about half the volume of water which would be required for its excretion in the urine.

The nasal gland has been shown to have a salt-secreting function in the cormorant, herring gull, and the Humboldt penguin (*Spheniscus*), and such a function is inferred from the gland's gross and microscopic anatomy in the pelican (*Pelecanus*), eider duck (*Somateria*), and petrel (*Oceanodroma*). Within a single genus such as the gull (*Larus*), the size of the gland increases with the extremeness of marine habitat.

Apropos of this remarkable adaptation in the marine birds, it may be noted that, unlike the elasmobranch fishes and the Jurassic ichthyosaurian reptiles, no ovoviviparous bird is known among the recent or fossil birds (though the avian fossil record is admittedly one of the most incomplete chapters in paleontology, being limited to two fossils!).

Notable among the truly marine birds is the albatross, of which two species (the Laysan albatross, *Diomedea immutabilis,* and the Black-footed albatross, *D. nigripes*) nest on Midway and Laysan Islands. The albatross was dubbed the Gooney Bird by sailors because of its presumed lack of intelligence; in captivity, however, it proves to be intelligent, adaptable and easily trained, the contrary impression probably stemming from its awkward gait, mischances in landing from flight, and extreme docility. It is the largest and strongest of all sea birds, and early explorers of the South Seas were cheered by its companionship, a bird often accompanying a ship for days without alighting on the water and, by tradition, even sleeping on the wing. The albatross was considered a good omen, and the evil fate of him who shot one with the cross-bow is familiar to readers of Coleridge's *Rime of the Ancient Mariner.*

When about eight or nine months old the albatross abandons island life for the open sea, where it may remain for five or even seven years, subsisting entirely on fish or squid. Its paired nasal glands are large and located in bony sockets above the eyes, and discharge their secretion through small openings beneath the tube-nostril, whence the fluid drains along grooves in the beak to drip off the end. When actively secreting, drops of fluid may drip from the beak at 2- to 10-second intervals, and sometimes the bird shakes the drops off the end of the beak.

The nasal gland of the albatross, like that of other marine birds, has been known to ornithologists for years but its true function was unrecognized until 1957, when, under the auspices of the Office of Naval Research, Drs. Hubert and Mable Frings, of Pennsylvania State University, visited Midway Island for the purpose of exploring means of repelling these (and other) birds from the runways used by jet and other planes. (One albatross in a jet air-intake or propeller can spell catastrophe for more than the bird.) Familiar with the work of Knut Schmidt-Nielsen and his co-workers on other marine birds, the Frings carefully observed the albatross and concluded that here also the nasal gland is an organ related to the maintenance of salt balance. They later showed that the nasal secretion consists of sodium and potassium chloride, the concentration of these cations averaging 829 and 24 milliequivalents per liter, a mixture having almost twice the osmotic pressure of sea water, three and one-half times that of the blood. Since nasal dripping was observed to occur when the birds had been fighting with each other, during their ritual dancing, or even during the excitement of feeding time, it was inferred that the nervous control of the gland was such that during moments of stress excess secretion might actually lead to salt deficiency.

The albatross had never previously been kept alive in captivity for more than a few weeks or months at most, the captive birds mysteriously dying without obvious disease. The Frings, however, shipped specimens of both species from Midway to the States and, following the lead suggested by the salt gland, kept them in good health: where others had given the birds fresh water to drink, the Frings gave them only sea water; and their diet of fish and meat scraps was supplemented with excess salt contained in either gelatin capsules or commercial tablets (0.8 gm.) every two or three days. If there was evidence of salt deficiency, salt was administered in amounts sufficient to cause abundant nasal dripping, which is indicative of excess salt in the body. It seems impossible to overfeed salt, so effective is the protection afforded by the gland, but without sea water to drink and (in captivity) excess salt in the diet, the loss of salt in consequence of what is perhaps incidental excitation of the gland leads to lassitude, weakness, coma, and death—a syndrome typical of salt deficiency. Inasmuch as water is retained in the body only *pari passu* with sodium, the primary disturbance in salt deficiency is probably excessive reduction of the volume of the body fluids (chiefly the plasma and interstitial fluid), rather than an imbalance between sodium and potassium or other salts.

In the cormorant, secretion has never been observed except after an osmotic load, and with no incidental stimulus to salt loss the bird is therefore not dependent on sea water ingestion. In the albatross, however, either because secretion by the gland is readily induced by stimuli other than sodium excess, or because of the habit of spending so many years at sea, the bird appears to be dependent on the ingestion of salt in excess of that contained in its diet of fish, etc., and this excess is available to it, of course, in its natural habitat only in the form of sea water. If other marine birds habitually drink sea water they are probably such forms as the petrel and penguin which feed on invertebrates such as squid, etc., which are osmotically equivalent to sea water and hence afford little derived water during metabolism.

The success of the Frings' venture is attested by the fact that at the time of writing four out of five alba-

trosses brought to the States in April, 1958, have survived captivity for 15 months and are now in the National Zoological Park, Washington, D.C. One bird died
of heat prostration after one month in Washington.

The discovery of the salt gland in marine birds immediately raised the question whether an homologous structure existed in the marine reptiles. Unlike the bird kidney, the reptilian kidney cannot concentrate the urine
at all above the osmotic pressure of the blood, and consequently such reptiles as do not have a source of water
available to them are in an even more precarious position with respect to salt and water balance than are the
birds.

Reptiles with a marine habitat occur in four different
orders: turtles, crocodiles, snakes and lizards. Five
species of turtles are strictly marine, returning to land
only to lay their eggs. At sea they subsist on fish, which
have a relatively low salt content, and on invertebrates
and seaweed, which are isosmotic with sea water. The
salt-water crocodile *Crocodylus porosus* is generally es-
tuarine, but it wanders far out into the Indian and
Pacific Oceans, living mainly on fish. The sea-snakes,
Hydrophidae, which inhabit the Indian Ocean, are also
fish eaters; the primitive *Lacticauda* lay their eggs on
land, but the more specialized sea-snakes are ovoviviparous. From the present point of view the most interesting of the marine reptiles is the Galapagos sea
iguana (*Amblyrhynchus cristatus*) which lives in the
surf-splashed rocks of the Galapagos Islands and sub-
sists on seaweed, with a large salt excess and water
deficit.

Schmidt-Nielsen and Fänge have demonstrated the
nasal secretion of a concentrated sodium chloride solution after the injection of hypertonic salt solution in the
brackish-water terrapin (*Malacolemys terrapin*) and the
completely marine loggerhead turtle (*Caretta caretta*),
and from the anatomy of the nasal gland infer that the

organ has this function in the green turtle (*Chelonia mydas*), the salt-water crocodile, in several sea-snakes (the gland seems to be absent in one sea-snake, *Pelamis*), and the marine iguana. Only in the iguana, however, is the drinking of sea water perhaps habitual because of the animal's diet of seaweed.

The allusions in Carroll's *Alice in Wonderland:* 'So they went up to the Mock Turtle, who looked at them with large eyes full of tears'; and again in Kipling's *Just So Stories:* 'for I am the Crocodile, and he wept crocodile-tears to show it was quite true,' and elsewhere to a propensity for tear production in reptiles failed to stir the scientific imagination, Schmidt-Nielsen and Fänge note, and consequently an important chapter in the regulation of the composition and volume of the body fluids has long been neglected. Probably neither Carroll nor Kipling ever saw a marine turtle or crocodile cry. and if they had they could not have known how salty the tears would be. But the phenomenon was possibly well known to seafaring men of their day, an incredible strand of verity entangled in unreliable legend—for even in Carroll's day, mock-turtle soup was made with calf's head, veal or other meat.

With the possible exception of the iguana, the salt gland in the marine reptiles, as in the birds, probably serves as a safety device, protecting against excessive ingestion of salt or loss of water. No such organ is present in the marine seals or whales, which by present evidences must rely on the kidney for body fluid regulation. The discovery of the salt gland in the marine reptiles will perhaps send the paleontologist scurrying back to the fossil record because it may be possible to judge the relative size of the gland in better preserved skulls and to determine if this unique adaptation aided the Triassic and Jurassic ichthyosaurs and plesiosaurs and the Cretaceous mosasaurs to establish themselves in a marine habitat.

      •   •   •

It is well known that men lost at sea cannot drink sea water and survive, and that to do so only shortens life. With the maximal osmotic concentration of human urine set at a level only slightly above that of sea water, very little of the water in the latter could even theoretically be made available for the excretion of other urinary constituents. About 500 cc. of sea water per day is about all that can be tolerated without gastrointestinal disturbances from the unabsorbable magnesium and sulfate. This would yield a meager 143 cc. of free water, a negligible quantity in the face of the minimal requirement of 500 cc. for urine formation, and upwards of 1500 cc. for sweat under conditions of exposure to sun and wind. Consumption of larger amounts would only lead to diarrhea and further dehydration, and thus accelerate catastrophe.

Nor can a man lost at sea improve his position by drinking urine; his kidneys are already doing the best they can and the only use to which either urine or sea water can be put is to use them to cool his clothing by evaporation, thereby cutting down on water loss in sweat. Taking the minimal urine volume as about 10 cc. per gram of protein metabolized, eating raw fish will not ameliorate dehydration, since all the water available in the fish is required to excrete the protein metabolites. A man might gain some water by drinking the juice expressed from fish muscle, since this is largely free of protein, but no one has yet been able to devise an effective method of doing this in a small boat at sea because the juices of the muscle are most tenaciously held. Man simply cannot venture out upon the sea or into the desert except in the security afforded by his providence.

CHAPTER XII

MAN

'Man was certainly not the goal of evolution, which evidently had no goal. He was not planned, in an operation wholly planless. He is not the ultimate in a single constant trend toward higher things, in a history of life with innumerable trends, none of them constant, and some toward the lower rather than the higher. Is his place in nature, then, that of a mere accident without significance? . . . The situation is as badly misrepresented and the lesson as poorly learned when man is considered nothing but an accident as when he is considered as the destined crown of creation. His rise was neither insignificant nor inevitable. Man *did* originate after a tremendously long sequence of events in which both chance and orientation played a part. Not all the chance favored his appearance, none *might* have, but enough did. Not all the orientation was in his direction, it did not lead unerringly human-ward, but some of it came his way. The result *is* the most highly endowed organization of matter that has yet appeared on the earth—and we certainly have no good reason to believe there is any higher in the universe. To think that this result is insignificant would be unworthy of the high endowment, which includes among its riches a sense of values.'

Thus writes George Gaylord Simpson in *The Meaning*

of Evolution, a recent Terry Foundation Lecture on 'Religion in the Light of Science and Philosophy.'

If this definition of man is acceptable to the reader, we may inquire in greater detail how man came into his superior endowments.

Man is first of all a placental mammal; secondly, a primate among mammals; then a primate uniquely characterized by the fact that he walks entirely on two legs, leaving the forelimbs free for use as hands; and lastly, by virtue of the great development of his brain, he is an acutely conscious, and very self-conscious, creature—capable, to a degree immeasurably greater than any other animal, of profiting by individual and social experience.

Back in the Cretaceous his ancestor was a small insectivorous animal related to the tree shrews which survive today only in Borneo. This animal had five digits on each foot with which it scratched, dug, grasped and climbed; it sought its food in the trees as well as on the ground, and in addition to eating insects it probably relished berries, birds' eggs, and nestlings, and when feeding it sat up on its haunches in the manner of a squirrel, clutching its food in its front paws. It probably did most of its hunting by the sense of smell, and depended on smell and hearing to escape from its enemies, as did all other animals of the Mesozoic era.

In the Lower Eocene this insectivore had produced the lemuroids, the lowest animals in the order Primates, the order which includes, among living forms, the lemurs, marmosets, monkeys, apes, and man. Highly modified representatives of the lemuroids survive in the lemurs, galagos, and lorises of Madagascar, Africa, and southeastern Asia, respectively. By the end of the Eocene the lemuroid stem had given rise to the tarsioids, of which the spectral tarsier, *Tarsius*, of Borneo, is the sole surviving remnant. The noteworthy difference between the lemurs and tarsioids is that in the former the eyes look to either side, while in the latter they look directly forward so that the visual fields overlap and, by rotating the head, an object can be scanned by the eyes in alternation and the animal can thereby obtain increased perception of distance and dimension. The living *Tarsius* has utilized this rotary motion to such an extent that it can turn its head through an angle of almost 180° and thus look directly behind itself. But in whatever direction it looks, the two eyes engage in parallel rather than convergent vision.

Below the primates the brain had been predominantly concerned with smell, while vision, hearing and touch had been merely auxiliary senses. But when the lemuroids and tarsioids took to living in the trees, smell lost most of its usefulness and the other senses became increasingly important. This shift in importance is reflected in the structure of the brain: as compared with other mammals, in the primates the areas concerned with vision, hearing, and touch are greatly enlarged relative to the olfactory area—with a proportional increase in those areas concerned with muscle sense and muscular activity, which are functionally related to these senses.

In the late Eocene or early Oligocene the tarsioid stock gave rise to the monkeys, in which the two eyes are subject to conjugate movements and can be converged upon a nearby object, producing true stereoscopic vision with its wealth of detail concerning distance, size, contour, quality, and solidity. Vision, touch and hearing now became the dominant senses, while the archaic sense of smell was relegated to an inferior position. Some of the monkeys of the New World had larger brains per unit of body weight (one to seventeen) than man (one to thirty-five), but they also had prehensile tails and they used their brains only to stay up in the trees, having no good reason to do otherwise. But the monkeys in the Old World, including the surviving macaques, baboons, and mandrills of southern Asia and Africa, never had a prehensile tail—indeed, they had no tail at all to speak of

—but boasted instead, where they sat down, callosities
that were frequently surrounded by red, blue or purple
skin and that became enlarged in the female just before
ovulation, an event which was followed, for the first time
among mammals, by external menstrual bleeding. The
absence of a prehensile tail in the progenitors of the Old
World monkeys may or may not be related to the fact
that among them there developed the tailless apes that
lived for the most part on the ground, notably the genus
Dryopithecus (tree ape) which roamed over Europe,
Africa and Asia in the Miocene and, as far as teeth go,
could have given rise to the gibbon, orang, chimpanzee,
and gorilla, or to man

The circumstances surrounding ape-to-man evolution
are still obscure, but all authorities are agreed that the
step is importantly concerned with the assumption of
the erect attitude and the liberation of the forelimbs for
use as hands, a transition to which both the human brain
and skeleton bear ample witness. The great apes in varying degree use their forelimbs to swing from branches,
and on the ground they more or less support the body
on the knuckles. The transition from the brachiating to
the bipedal habitus might be expected to occur under
circumstances where forms that had hitherto been
arboreal were forced by recession of the forests to seek
their food in the open plains. In assuming the upright
posture, the foot had to be remodeled in order to balance the body; then the pelvis, backbone, and head had
to be realigned until the arms hung free at the sides.
Once the hands were relieved of all responsibility in locomotion, natural selection could foster the further evolution of the brain around the sense of touch and the motor
activities involved in manual dexterity.

Recognized human remains do not go back of the
Middle or Early Pleistocene, at most not more than one
million years ago. The preceding Pliocene saw the beginning of the most recent of mountain-building episodes, the Cascadian disturbance which, continuing

through the Pleistocene, culminated in the uplift to their
present altitudes of the Himalaya, Andes, Rocky Mountains, and the Alps. As this disturbance raised Eurasia
and buckled its ancient, eroded mountain remnants into
new and jagged peaks, that continent, in common with
other areas in the Northern Hemisphere, was subjected
to refrigeration presaging the glaciation of the Pleistocene. In that part of Central Europe lying between the
Scandinavian ice to the north and the Alpine glaciers to
the south, the mean annual temperature fell below the
freezing point, while China and India, except in the extreme south, were probably not much warmer. Although
the summers may have had a few short warm spells, the
winters were severe. As the climate became less hospi-
table and the forests dwindled in size in the late Pliocene,
the dryopithecine apes came to an end in Eurasia while
their cousins continued to flourish in the African forests,
which at that time extended northward to the shores of
the Mediterranean. Darwin held that Africa was probably the home of the ancestors of man, and subsequent
discoveries have fully supported this belief, though the
available information throws no light on whether the
critical phase of his evolution transpired in the forests or
the plains.

The evolutionary 'missing link' between the apes and
man was for long so conspicuous by its absence that it
became a subject for jokesters who had a gap of their
own to fill. The biologist is now at no disadvantage in
respect to the humorist's ancestry, or his own. The family
tree begins somewhere among the dryopithecine apes of
the Pliocene, some 5,000,000 to 10,000,000 years ago—
so far back in time that a detailed search of the sparse
fossil record would be of little moment here. From a
common dryopithecine stock there evolved two biological families, the Pongidae or anthropoid apes, and the
Hominidae or manlike creatures. The close similarity
between the Pongidae and modern man led Darwin's
friend, Thomas Henry Huxley, to conclude that they had

had a common evolutionary origin, a view first set forth in Huxley's famous essay entitled *Man's Place in Nature*, published in 1863. The biological affinity between the anthropoid apes and recent (as well as fossil) man is so close that the Pongidae and Hominidae are today included in a superfamily, the Hominoidea.

Of the Hominidae, only one species—*Homo sapiens*, or modern man, survives. In Huxley's time indubitable human or related fossils were almost unknown and the few suspected fragments could not be confidently classified. In the past century, however, these fossil fragments have been unearthed in such numbers as to keep the paleontologists and anthropologists busy describing them, naming them, and generally hotly debating their interrelations. But always these fossils have been either man or ape, never man-ape—and the biologist has been too acutely conscious of the importance of the 'missing link' to find the subject very funny.

Then in 1925 R. A. Dart discovered at Taungs, in Bechuanaland, a fossil skull of a new 'anthropoid' which he named *Australopithecus africanus* (African southern ape), and which he suggested was closely related to the human stem. (See Figure 11.) Since that date large quantities of the fossilized remains of *Australopithecus* have been collected from cave deposits at widely separated sites in the Transvaal. These fossils have precipitated a lively debate, some paleo-anthropologists wholly denying their human affinity, and contending them to be the remains of advanced apelike forms closely allied to the gorilla and chimpanzee. The most extreme opposition to Dart's interpretation has come from a minority who believe that man was descended, not from an early anthropoid-ape stem but from an unspecialized monkey-like quadruped which had not yet acquired the brachiating (arm-swinging) habitus and other anthropoid features, and who sought (with scant warrant) to place the separation of the hominid (or distinctly humanlike)

stem from the other primates as far back as the
Oligocene, perhaps 30,000,000 years ago.

One major difficulty in determining the proper evolutionary position of the South African forms is the un-
certainty of dating fossils or associated materials when
the age ranges from, say, 100,000 to 1,000,000 years or
more, because within this range no good chemical or
isotopic method is yet shown to be applicable. If the
australopiths were evolved during the Middle Pleistocene (ca. 500,000 years), and if others of the accepted
homininds are this old or older, then the australopiths
could not be ancestral to man, no matter what their
anatomical features, because, as Simpson has observed,
one of the basic laws of biology is that a man cannot be
ancestral to his grandfather. However, the dating problem, though still on a relative rather than absolute scale,
has now improved to the point where Dart's interpretation can be accepted without offense to Simpson's 'law.'
The earliest australopithecine remains so far known
(Early Pleistocene) apparently antedate the oldest
member of the Hominidae, *Pithecanthropus* (Middle
Pleistocene) by some hundreds of thousands of years.
Though no one of the several local variants of the genus
Australopithecus now known from South Africa may
have been directly ancestral to the genus *Pithecanthropus*, and though both genera may have coexisted
at some interval, the two types at least present a proper
sequence in time.

Anatomically, *Australopithecus* conforms so closely,
in multitudinous and highly distinctive details, to the
requirements for the connecting link between the anthropoid-ape stem and *Pithecanthropus* that a true ancestral relationship seems extremely probable. In recent
years the tendency to erect a new species or even genus
on every jawbone or cranial fragment has given way to
taxonomic consolidation, owing, in part, to increased
recognition of the reality and importance of the anatomical variation observable in any one species (such,

for example, as *Homo sapiens*). Thus the genus *Australo-pithecus* is now taken to include both Broom's *Plesian-thropus* (*plesios* = near; *anthropos* = man) and Robin-son's *Telanthropus* (*teleos* = perfect; *anthropos* = man), so that one is tempted to wonder if Dart might not have done better to combine *Australo* (southern) with *anthropos* (man) rather than with *pithecus* (ape). *Aus-tralopithecus* was truly hominid, perhaps the earliest truly hominid (if a sharp line can be drawn), in that in many respects he qualifies as the precursor of all later hominid forms. His skull, pelvis, and femur show that he walked upright in the bipedal manner, and it may be presumed that he used the forefeet as hands. There is some evidence that he knew the use of fire and even a rough stone weapon, but here the evidence ends. But anatomically he affords the starting point for the almost insensible succession of the later hominids, *Pithe-canthropus* (which includes *P. erectus* of Java, *Sinan-thropus* or *P. pekinensis* of China, *Atlanthropus* of Algeria, and possibly Heidelberg man), this genus ex-tending back 200,000-500,000 years; then Pre-Mouste-rian man (100,000-200,000 years), Early Mousterian man (50,000-100,000 years); then *Homo neanderthal-ensis* (ca. 50,000 years) and modern man *Homo sapiens* (ca. 40,000 years). Extreme taxonomic reduction would put Mousterian man and Neanderthal in the genus *Homo*, if not in the species *sapiens*.

Otherwise *Australopithecus* was an ape, so distinctly simian in respect to teeth, jaws and brain case that his close affinity with the common stock from which stems the later Pongidae is indubitable, and Dart's name is justified. Perhaps he goes back 1,000,000 years, to the opening of the Pleistocene. His immediate pongid an-cestry is unknown, but is to be sought among the primi-tive apes which, descended from *Dryopithecus* of the Miocene, ranged over wide areas of the Old World in Pliocene time. A single fossil primate, *Oreopithecus*, has been recovered from Early Pliocene deposits in Italy; the teeth have a distinctly hominid character, but the claim that *Oreopithecus* affords a link in the direct ancestry of the Hominidae is not as yet supported by other anatomical data and is contrary to the evidence that the hominid stock was not evolved until some millions of years later.

It is not possible to determine the level of intelligence attained by these extinct forms except as we write a rough equation between intelligence and brain weight per unit of body weight. This equation is of little use, if for no other reason than because the brain's potentialities, in the chimpanzee or in man, invariably exceed its actual attainments. The chimpanzee, with its remarkable capacity for learning and solving problems, is handicapped by the absence of articulate speech, by which sounds can be converted into symbols and rearranged to give new meanings, and without speech it is impossible to judge how bright, by human standards, a chimpanzee is. On the other hand, untutored man remains an intellectual savage—it is the cultural treasure imparted by education that converts him into a creator of science, literature, music, philosophy, and art. The anthropologist infers that the mental capacity of fossil man increased generally in the phylectic sequence from *Australopithecus* to *Homo sapiens*, and that the first was probably very near the beginning of speech. Probably also in *Australopithecus*, ratiocination, the debating of causes and effects, had begun to play a more important role than instinct.

Many of the important features in the transition from ape to man represent what the biologist calls paedogenesis (*pais, paidos* = child; *gignesthai* = to be born), a phenomenon frequently observed in evolution and representing the progressive arrest of development of a new species in the juvenile or fetal state of the ancestral form. Although at birth the human infant is by far the largest newborn among the primates, it is, nevertheless, the

most 'fetalized' in many anatomical features: its development is retarded in respect to teeth, the complete absence of hair from the body, the prolongation of the
period of gestation, the slower closure of the sutures of
the skull, delayed sexual maturity, and delayed matura-
tion of the brain. Of all fetal characters, the late closure
of the cranial sutures and the postnatal development of
the brain are of the greatest importance. The human
infant is extraordinarily helpless at birth; its brain not
only lacks instinctual patterns which might enable it to
be self-sufficient, but it is so immature that it is deficient
in learning capacity. Hence survival depends not only
on a long period of maternal care but on an equally long
infancy and childhood spent in the family and com-
munity. The child's earliest acquired knowledge is chiefly
of the type that is imparted by parental tutelage, rich
in the complex and sophisticated culture of the social
group. It is partly because man's brain is undeveloped
at birth and almost utterly devoid of instinctual patterns
that it has the potential capacity to develop with maturity into the wonder-working organ that it is.

When Claude Bernard first developed his idea of the
internal environment in 1857, his attention was fixed on
the specific properties of the body fluids. But twenty
years later his thought had shifted to the organism as
a whole, and in his Lessons on the Phenomena of Life
Common to Animals and Plants (1878–1879), he points
out that the higher organism is so constituted that when
it is disturbed it reacts in such a manner as to restore
the balance: 'All the vital mechanisms, however varied
they may be, have only one object, that of preserving
the conditions of life in the internal environment.'

If this epitome requires amendment, it is only with
respect to the preposition *in*. When we ask 'What is the
object of all the vital mechanisms?' we must reply that
it is certainly not just the constancy of the internal environment, which is a lifeless solution and only one of

several means by which the organism seeks to achieve a free and independent life. All the vital mechanisms operate to preserve the constancy of the internal environment only because this environment is the last ditch of defense between the living cells that comprise the organism and a hostile world. All the vital mechanisms operate to preserve 'the conditions of life' *through* rather than *in* the internal environment.

This idea was well expressed by the Belgian physiologist, Léon Fredericq, who, in 1885, declared: 'The living being is an agency of such a sort that each disturbing influence induces by itself the calling forth of compensatory activity to neutralize or repair the disturbance. The higher in the scale of living beings, the more numerous, the more perfect and the more complicated do these regulatory agencies become. They tend to free the organism completely from the unfavorable influences and changes occurring in the environment.'

The heart of the problem thus lies not in the internal environment but in the 'organism' that has created this hothouse for itself; which means that the heart of the problem lies basically in the structure of protoplasm, the living substance out of which all organisms are composed. The living unit in all plants and animals is a microscopic bit of protoplasm called a 'cell,' characterized by sharply limited physical dimensions and sharply limited capacities for physiological reaction, by differentiation into nucleus, cytoplasm, cell membrane, and other cellular apparatus, by the capacity for cell division and hence for reproduction, and by genetic continuity from one generation of cells to another through the genes and chromosomes.

It is axiomatic in physiology that this protoplasmic cell never acts—it only reacts; and it reacts only to restore the internal *status quo:* protoplasm is basically a physical-chemical mechanism having the character of a self-integrating, self-restoring, self-centered comfort machine, operating to the sole end of self-preservation: that is why and how protoplasm, or some biotic precursor of protoplasm as we know it, came into existence, survived, and multiplied in the Archeozoic seas to produce the first living cells—once called the Protozoa (*protos* = first; *zoion* = animal) but better called Protista (from *protistos*, meaning simply 'first'), because at this level it is meaningless to distinguish between animals and plants.

When, at a higher stage of evolution and in consequence of natural selection, the products of cell division formed a lifelong association to give rise to the multicellular animals or Metazoa (*meta* = between; *zoion* = animal), certain cells became specialized in such a way that physiological labor might be divided between them: muscle cells to contract, gland cells to secrete, bone cells to give physical support, nerve cells to conduct, while the reproductive cells remained protected in the germinal tract to reproduce generations without end. It is estimated that the body of a man weighing 155 pounds contains 26,500,000,000,000 cells (not counting the red blood cells, of which there are 10,240,000,000,-000 in each of his 12 pints of blood). Except for the subtle dynamics of cell reacting on cell, nothing intrinsically new was added in the Metazoa: though not so elaborately or quickly, the protistic cell can contract, conduct, secrete, resist, divide; it possesses every intrinsic physiological property discoverable at the metazoan level.

The metazoan is thus a congeries of many cells in unsteady balance with each other, incessantly disturbed by the internal stresses of metabolism, growth, and deformation, as well as by a multitude of external stimuli; and by a wide variety of devices seeking to reduce these stresses to a minimum, just as did the protistic cell. In the evolution of the kidney, the organism acquired an organ that operates automatically to supply the multitudinous cells of the body with an immutable environment in which to live with minimal disquietude. The evolution of the nervous system represents the evolution

of a device that, by introducing flexibility and adapta-
bility into the stimulus-response pattern, supplements
the operations of the kidney to this same end—to minimize the organism's disquietudes.

The role which the nervous system plays in the perception of various sensations, in the integration of perceptions, both past and present, and in eliciting motor
activity, both voluntary and involuntary, is obvious to
everyone who knows the meaning of 'brain' and 'nerve.'
But the role of the nervous system as viewed in the
long perspectives of evolutionary history can bear a word
of elaboration.

Despite its apparent complexity, the nervous system
involves only four basic operations: first, the conduction
of a wave of excitation (actually a reversal of the
physical-chemical polarization of the protoplasmic surface) over the undifferentiated protoplasm of the cell;
second, the more rapid conduction of a wholly similar
wave of excitation over the long processes (axons) of the
specialized nerve fiber; third, the secretion of humoral
or chemical agents that serve to excite or inhibit other
cells, including other nerve cells; and fourth, transmission across the 'synaptic' junction between nerve and
nerve—and here it must be emphasized that nerve fibers
are not connected together anatomically, any more than
are other cells, but nerve meets nerve across a sort of
physiological spark gap which the physiologist calls the
'synapse' (*synapsis* = conjunction, union).

A specialized cell similar to the nerve fiber first makes
its appearance in certain Protozoa; in the lower Metazoa, nerve fibers are combined into a nerve net which
permits rhythmic activation of successive segments of
the body, as in the progressive movements of the legs of
the centipede.

Humoral or chemical excitation and inhibition, effected by such substances as acetylcholine and adrena-
line, are widespread among the invertebrates and play

an important role among the vertebrates in the transmission of nerve impulses across the synaptic junction
and also across the junctions between nerves and other
cells, such as those of muscles and glands.

The synaptic junction between nerve and nerve, first
appearing in the jellyfish and allied forms, introduces
the unique features of one-way conduction and pro-
visional or conditional response. Nerve impulses arriving
simultaneously at a synapse may have an additive effect,
thus producing locally the phenomenon which the
neurophysiologist calls 'summation'; or, alternatively,
such convergent impulses may prevent transmission from
one nerve to another, producing synaptic 'inhibition.'
The synaptic junction has been likened to a valve: it is
the essential mechanism in the reflex arc which subordi-
nates the activity of the arc to events occurring elsewhere in the body.

The nervous system acquires its capacity for complex
reactions by combining these four basic operations—
protoplasmic conduction, neural conduction, humoral
excitation or inhibition, and synaptic conduction—into an
elaborate network of communication extending to all
parts of the body, and co-ordinated by a 'central nervous system.' In the lower metameric animals (take the
centipede again, or the angleworm) the central nervous
system consists of a chain of reflexes in which each event
in the chain gives rise to impulses eliciting the next reaction in the series, the entire series being 'fired' by, or
under the control of, one or more large nerve centers or
ganglia in the anterior end of the body.

In the protovertebrate the 'central nervous system'
was probably a simple longitudinal tube of nervous tissue
with which the peripheral nerves were connected in a
segmental fashion, to meet the needs of the segmentally
arranged muscles. This is essentially the pattern of the
spinal cord in all vertebrates. But as the animal increased
in complexity, and particularly with the progressive evolution of the distance receptors—the nose, eyes, and ears

—the reflex centers related to these organs and located
in the anterior end of the spinal cord became enlarged
to give rise to massive ganglia that took the form of a
'nose brain,' an 'eye brain,' and an 'ear brain,' which
supplemented older ganglia—representing a 'visceral
brain' concerned with the gills, stomach, and other internal organs, and a 'skin brain,' concerned with the
elaborate 'lateral line' and other sense organs which are
present in the skin of fishes. Because of the linear arrangement and segmental distribution of these master
ganglia, this part of the brain is generally designated as
the 'brain stem'; and, because these ganglia were fully
developed, in the qualitative sense, in the fishes, the
brain stem is sometimes called the 'old brain.'

As the early vertebrates engaged to an increasing extent in crawling, swimming and other movements, another ganglion of the central nervous system developed
to form the cerebellum, a part of the brain which is
concerned chiefly with motor co-ordination and the
orientation of the body and its appendages in space, and
the functions of which in all animals remain wholly
reflex and unconscious. The relationships between the
ganglia of the brain stem and the distance receptors are
beautifully exhibited in the 'linear' brain of the common
dogfish, which is studied by every student of comparative anatomy. But the over-all relations of the brain stem
and cerebellum have been reconstructed from well-
preserved fossil casts in the Devonian ostracoderm
Cephalaspis and found to have a pattern not very dissimilar from that of the living lamprey, so that we may
conclude that the 'old brain' or brain stem had been
shaped in its fundamental features as early as the
Silurian, and perhaps the Ordovician period.

The progressive enlargement of the central nervous
system in the anterior end of the body is called 'en-
cephalization' (*en* = in; *kephale* = head), meaning, in
common parlance, the development of a 'brain in the
head,' whence the brain is sometimes called the 'enceph-

alon.' Dominance by the cephalic portion of the nervous system is common among the invertebrates (flatworms, annelid worms, mollusks, insects, and others), but in no instance has such a complex brain been developed as in the vertebrates. The evolution of the vertebrate brain has its roots in the fact that the protovertebrate was a spindle-shaped creature that, when it essayed to enter the fast-moving waters of the Cambrian continents, swam vigorously with one end foremost and needed to concentrate not only the distance receptors, but also the motor control of the mouth parts, the respiratory mechanism, and the muscles generally, in the front end of the body.

This brain stem served the ostracoderms and fishes, and, with but slight elaboration, the Amphibia, reptiles, and birds, through geologic ages, only to undergo dramatic evolution in the mammals. In all animals below the mammals, the most anterior part of the brain consists of two nearly separated lobes (cerebral hemispheres) which primarily subserve the function of smell (the old 'nose brain'). In the reptiles, the cerebral hemispheres have begun to enlarge and to develop, for the first time, an outer layer or cortex composed of relatively large pyramid-shaped cells having many interconnections with each other, as well as with the brain stem. As between the lowest mammals, such as the duckbill and echidna, and man, the cerebral cortex undergoes enormous increase in relative size, partly by increase in the number of cell layers and partly by wrinkling or infolding of the brain substance so as to increase the surface area. This increase in number of cells in the cortex is paralleled by a corresponding increase in the number of nervous interconnections enjoyed by each cell, and by the development of large association tracts that connect all parts of the enlarging brain together so that all regions, however remote from each other, are interconnected. So great is the relative development of the cere-

bral hemispheres in the anthropoid apes and man that
they cover and almost envelop the brain stem and constitute the preponderant mass of the brain as a whole.

But in all this multiplying of nerve cells and compounding of the number of their potential interconnections in the evolution of the brain stem and cerebral cortex, no new functional features have been added, so far
as can be determined, to the basic patents that go back
to the very roots of the evolution of the nervous system.
As the nervous system has increased in complexity, behavior has evolved from the simple to the complex simply by grafting the new on to the old, so that the central
nervous system of the mammals—the cerebral cortex,
brain stem and spinal cord—is in essence a hierarchical
system which works the way it does by virtue of the
fact that the new can dominate the old.

Many of the activities of the body—circulation, digestion, secretion, excretion, body-temperature control—are
taken care of by the autonomic or 'vegetative' nervous
system, and normally these activities have no counterpart in consciousness, so that a man remains wholly un-
aware of them. But neither is a man aware of the in-
finitude of automatic operations involved in deliberate
or voluntary action. In such a seemingly simple act as
walking, hundreds of thousands of neural pathways
make their contribution to each step: sensory impulses
from the soles of the feet, the joints and tendons of the
ankles, legs, thighs, back, arms, and neck, and still others
from the semicircular canals of the middle ear as well
as visual impulses from the eyes, converge on the brain
stem and cerebellum to effect the perfect co-ordination
of an equally large number of motor impulses activating
millions of muscle fibers. Locomotion is in fact a highly
complicated performance that is achieved smoothly and
easily only because the organism does not have to think
about the multitudinous details. As pointedly phrased by
Mrs. Edward Craster:

> *The centipede was happy quite*
> *Until a toad in fun*
> *Said, "Pray, which leg goes after which?"*
> *That worked her mind to such a pitch,*
> *She lay distracted in a ditch,*
> *Considering how to run.*[1]

The difference between the toad and the centipede is that, rhetorically at least, the toad can think and ask questions. Among the notable features that distinguished the Pennsylvanian Amphibia from their cousins among the air-breathing fishes was an increased capacity to consider how and why to run. The eyes, now adapted for looking through air instead of water, could survey a greatly widened world across which the feet might carry them from here to there; and the brain could not only dictate the sequence of movements of the legs but it could integrate new sensory information into the total operation and, above all, retain a more complicated recollection and better forecast the consequences of going from here to there. But the emergence of the Amphibia from the water did not represent the 'emergence' of something qualitatively new in respect to brain function: the ostracoderms, the sharks, the air-breathing fishes before them, the reptiles, birds, and mammals after them, all had need of, and all possessed, some measure of awareness of themselves and their environment, and a corresponding ability to react accordingly.

A man can learn more than a toad, and a toad more than a centipede, because of differences in the degree of elaboration of the brain. In chemical composition, metabolic rate and anatomical structure, the brain of man, the toad, and the centipede do differ in details, but differences in learning capacity and in behavior must be sought not so much in these qualitative differences as in the sheer number of nerve cells and the potential variety of their interconnections. It is said that under

[1] *Pinafore Poems* 1871.

favorable conditions every animal can learn by the trial-and-error method. The one-celled Protozoa can improve through practice their ability to avoid unfavorable situations. A single arm of the starfish, a single tentacle of the sea anemone, can 'learn' independently of the rest of the animal. Clams and snails adjust themselves to new conditions, and the earthworm can learn in a few days whether to turn left or right when in search of comfort. Simple avoidance-reactions of one type or another are quickly acquired by the snail, squid, octopus, and cockroach, while the fiddler crab can be taught to move in the direction of its small claw, though its normal mode of progression is in the direction of the large one. The honeybee uses a sign language that involves 'symbolism,' and wasps and ants, like bees, show considerable plasticity in behavior. Fish, though not outstandingly versatile, can be conditioned to color, form, and sound, and the habit may be retained for several weeks, while carp are well known for their capacity for training in respect to feeding habits. Frogs and toads have a sense of place and show homing reactions; they can learn to find their way out of simple mazes and can remember the solution for periods of at least thirty days, and their breeding season takes on, for the first time in vertebrate history, the audible call of mate to mate.

Memory of places and persons is demonstrable in birds, and for over forty centuries man has been training the falcon to retrieve her living prey. It is believed by fanciers that the instinctive singing of the canary is improved by auditory tutelage under a Hartz mountain maestro, and canaries brought up with nightingales tend to copy the nightingale's song. It has been shown that pigeons can count up to five, ravens and parrots up to seven, by what might be called nonverbal thinking (that is, thinking that excludes verbal enumeration and the use of symbols) and this is about as good as man can do, and probably better than the chimpanzee. In respect to the ability to learn, however, the primates are superior to the infraprimates, while among the subhuman primates the chimpanzee is the most adept at problem-solving and the use of tools. Among the mammals generally, the carnivores belong above most rodents and other herbivores. Conversely, many lower forms can doubtless perceive relations of temperature and salinity (fishes), smell and taste (other mammals), vibration (bats), direction (birds), and other stimuli to which man is wholly insensitive because he does not have the appropriate receptor organs; but this is a matter of qualitative equipment. The basic fact is that the capacity to learn was perfected early in the course of evolution of the brain; it is not the fact of learning but *what* is learned that differentiates animals in the evolutionary scale, the higher animals being able to perceive relations that are beyond the comprehension of the lower forms.

In man, learning is importantly concerned with abstractions such as numbers, qualities, symbols and relations, and specifically with language, and, as Emerson at the age of twenty wrote in his journal (1824), 'Man is an animal that looks before and after.' By turning his attention from concrete objects to abstractions and to the past and to the future, he has welded his thinking processes into a mighty tool. But when a writer manipulates the twenty-six letters of the alphabet into a meaningful sentence, or the mathematician manipulates numbers from zero to infinity, he is using no specialized nerve cells and no unique cellular operations. In respect to the basic mechanism of operation, every nerve cell is like every other nerve cell, a generalization that applies not only to the different types of nerves in man but to human nerves, frog nerves and the nerves of the octopus. And every impulse is, except for intensity and duration, essentially identical with every other nerve impulse. Moreover, the structure of the nervous system is, on the whole, determined genetically and it is therefore predetermined at birth. No animal acquires or loses nerve cells, so far as is known, as a consequence of learning or experience,

or moves them anatomically from place to place; the in-
born pattern of the nervous system is the most rigid conceivable: a system of fixed point-to-point connections.
And yet the nervous system of the higher animals is the
most plastic and adaptable organ in the body.

In the rat, for example, the optic nerves from the right
eye are connected with the visual areas in the left lobe
of the cerebral cortex, so that a spot of light pinpointed
on the right retina is 'projected' (as the neurophysiologist
would say) on the left side of the brain with a spread of
no more than two or three cell diameters. Nevertheless
all but 2 per cent of the visual area in the rat's cortex
can be destroyed without affecting visual integrations,
as judged by learning reactions, and it need not be the
same 2 per cent that must be left for the animal to integrate effectively. Localization of certain functions is
much more evident in man than in the rat, but in both
species the network of nerve cells that makes up the brain
has properties of organization and responsivity that re-
side in activities of the system as a whole (or what is
left of it), rather than in its specific point-to-point connections. In learning, it is probably the pattern of activity
in the brain as a whole, or some large part of it, that
changes, and not the anatomical pathways.

In a crude analogy, the processes of mental activity
may be likened to messages spelled out on an electric
billboard: each bulb (meaning each nerve cell) flashes
on and off in a more or less invariant manner, but the
message is never twice the same and can be spelled out
with equal intelligibility by different bulbs and circuits
(that is, by different neurons in different areas of the
brain). The human cortex is estimated to contain ten
billion separate neurons coupled in three dimensions into
many, many, many times this number of potential circuits—taking even a fraction of the possibilities, the potential number of permutations and combinations is in-
comprehensible.

There is no difficulty in accounting for the plasticity

of the brain—the difficulty lies in explaining its lack of plasticity. Thomas R. Lounsbury, a nineteenth-century professor at Yale's Sheffield Scientific School, was not far wrong when he remarked that we must view with profound respect the infinite capacity of the human brain to resist the introduction of useful knowledge. Granting the force of this sardonic comment, by way of explanation it must be emphasized that the brain is basically just as conservative as any organ in the body: it works by repetition, and an organ that works by repetition can only learn by repetition—by being forcibly inscribed with new channels. Man differs from the other vertebrates in that he has, in some measure, learned how to learn.

CHAPTER XIII

CONSCIOUSNESS

The illusion of unchanging personality is probably common to all normal mammals with advanced brains. This illusion, as it appears to man, has long enjoyed the support of errors engendered by prescientific speculation, and notably by the dualism of René Descartes (1596–1650), the most eminent philosopher of seventeenth-century France. Descartes was initially interested in mathematics, to which subject he contributed analytical geometry and the familiar co-ordinates known by his name. Then mathematics led him into astronomy and to speculations about the origin and nature of the universe. Here he rejected the 'spirits' and 'genii,' to which even Kepler had assigned the movements of the heavenly bodies, in favor of an atomistic interpretation, and he bequeathed to physics the laws, first, that so long as a body is unaffected by extraneous forces it continues in the same state of motion or rest; and second, that simple or elementary motion is always in a straight line. He sought a natural explanation for gravity, heat, magnetism, and light; and, his own momentum carrying him in a straight line, he conceived that all chemistry, physiology, and biology could be explained in mechanical terms.

Descartes was as much concerned with expelling 'spirits' and 'genii' out of physiological phenomena as out of the sun, moon, and stars, and he considered man and animals to be, basically, very complicated 'machines.' Aristotle had sought to explain the differences between inanimate matter and plants by means of a 'vegetative soul,' between plants and animals by means of a 'sensitive soul,' and between animals and man by means of a 'rational soul.' Descartes rejected Aristotle's 'vegetative' and 'sensitive' souls, seeing in plants only the mechanical product of a developing seed, and in animals only curiously contrived machines that have neither consciousness nor feeling but 'act naturally and by springs, like a watch.' 'The greatest of all prejudices we have retained from our infancy,' he wrote, 'is that of believing that the beasts think.' Their 'life' is merely the beat of the heart, their 'feeling' merely the autoreaction of an organ, as when a plant moves to or away from the sun.

But, so it seemed to Descartes, man transcended animals by virtue of the fact that he was a thinking, rational being, and to explain this difference he retained in man's behalf Aristotle's 'rational soul,' coupling it to the mechanical body through the pineal gland, which was conveniently located close to the brain and for which there was no other demonstrated function.

Here, in the pineal gland, mind meets matter; here, receiving passively the data of the senses, it cogitates upon them; and here, in volition, it bends the body to its 'will.'

This Cartesian dichotomy between 'matter' and 'mind' lingers on not only in common parlance but in philosophy, giving rise to frequent discussions of the 'mind-brain problem' or the 'mind-body problem.' Some critical thinkers continue to adhere to dualism, holding to the belief in something called 'mind' which is other than matter, but the majority of workers in the biological sciences reject the belief in the existence of disembodied mind and see in 'mind' a mode or property of matter, so that psychical processes are wholly dependent on physical and chemical events in the nervous system. This philosophical position is commonly identified as 'materialism,' but the term 'naturalism' is to be preferred—because the ultimate nature of matter and energy remains unknown, and the task of science and philosophy is to study nature as given, and without prejudicial preconceptions.

Among the modern writers who have discussed the 'mind-body' problem at length is Sir Charles Scott Sherrington (1861–1952), to whom the world is indebted for the basic principles of neurophysiology, especially as related to the function of the central nervous system. Sherrington pointed out in his book, *Man on His Nature* (1941), that 'mind' is not equally an attribute of all living things. It is apparently absent in the molds, yeasts, fungi, and in the whole of the plant kingdom. Possibly present as some primordial awareness in the protozoa and lower metazoa, it makes its definitive appearance in the higher metazoa and reaches its maximal development in man. Yet even in such vertebrates as exhibit it in the adult stage, it is absent in the ovum and spermatozoon, and very meagerly developed, if at all, in the embryo. A newborn human infant cannot be said to have a 'mind' beyond the elementary perception of such things as hunger, discomfort, and fatigue. In every species, 'mind' appears only at a late stage in the development of the brain. Sherrington noted that the cells of the body are reproduced by cell division from the germinal cells, even as the germinal cells, carried in the reproductive organs, themselves divide and redivide, generations without end, so that the physical inheritance of the body goes back generation through generation into the geologic past. But there is no such continuity in respect to 'mind,' which, as it were, appears out of *nihil,* out of absolutely nothing, in each individual at a time when the brain has reached a critical stage of anatomic development; in other words 'mind,' unlike body, appears to be biologically discontinuous, or episodic. (Sherrington used the word 'phasic.')

Sherrington recognized—indeed, he himself went far to establish—that 'mind' is *utterly* dependent upon the most minute and detailed architecture and function of the brain for its realization, at least as something recognizable; and that it disappears, if not back into *nihil*, at least into utter unrecognizability with the disintegration of that brain. And yet, unable to see how 'mind' could be a manifestation of matter or energy, he was forced to conclude that it could never be examined as a form of either matter or energy.

One thinks, however, that Sherrington's difficulty arose in part from his persistence in the use of the word 'mind' in the Cartesian sense: as 'something that exists,' in the sense in which matter and energy exist. Long ago the philosopher David Hume defined 'mind' as an abstract term denoting the series of ideas, memories, and feelings which appear in consciousness, and which so overlap that they give the impression of being continuous, or of existing in a continuum. The lapse of two and a half centuries has only served to affirm Hume's definition, and there is no more warrant for speaking of this temporal sequence as a 'thing' than of so describing the *sequence* of flickering images that appear on the cinema or television screen. In the concrete sense, we can only speak of a particular image which momentarily appears upon the screen—that is, in consciousness. It would be well to drop the confusing word 'mind' from serious discussion.

But we can speak of 'consciousness,' either as subjectively perceived in ourselves or as inferred in others. Although there are plenty of common-sense tests for determining when a man is conscious or unconscious, an adequate, objective definition of consciousness presents considerable difficulty. Consciousness is not an abstract concept but a dynamic function of the organism in action, in the same sense as the heartbeat, respiration, muscular activity, or the excretion of urine. It cannot, however, be defined merely as the capacity to respond to stimulation, as is sometimes done, because an unconscious man can in some measure do this, as exhibited by reflex action, and so can a calculating machine to which we have no reason to attribute any measure of consciousness in the biological sense.

Stanley Cobb, the neurologist, defines consciousness as 'awareness of environment and of self.' But this definition is not wholly satisfactory because it fails to tell us how to obtain objective evidence of this awareness. Is a bird, singing in the dawn, a conscious creature? Or a fish, darting among the lights and shadows of the seaweed? Or an octopus, half-concealed in a rocky crevice, waiting for its prey?

It is well to defer attempting to answer these questions until we have found an objective definition of consciousness, a definition which we must seek in the broad perspectives of biology.

We may begin with man. As observed subjectively in ourselves, consciousness is variable in intensity, complexity, and duration. It comes and goes as between waking hours and sleep; it presents a distorted and fragmentary pattern in dreams and hallucinations; and, even in the waking state when it is normally continuous, it varies greatly in acuteness, reaching its peak in moments of excitement or intense attention when we are most 'wide awake.'

Abundant evidence in man as well as in other mammals indicates that consciousness depends on a state of continuous excitation of the cortex, and possibly requires for its presence a repetitive series of nerve impulses impinging on the cortex from lower centers. Loss of consciousness is sometimes associated with sharply circumscribed lesions in the brain stem, but these anatomical areas cannot be conceived to be 'centers of consciousness'; they are, rather, bottlenecks through which excitatory impulses between the brain stem and the cortex must pass in order to maintain the reverberating circuits.

Most men spend a third of their lives in sleep, a practically unconscious state, and yet the mechanism of sleep is poorly understood. Consciousness waxes and wanes like the aurora borealis as dreams come and go, shooting its loosened streamers into the dark recesses of the forgotten or the half-forgotten, and leaving such faint traces in memory that dreams are recalled with difficulty when one wakes up. Later we know that they are dreams, and sometimes we know that they are dreams even while we are dreaming, possibly because the flickering images are unreal and out of focus. Some neurologists speak of a 'sleep center' that is presumed to inhibit cortical activity, but with no greater cogency than others who speak of a 'wakefulness center' that excites cortical activity. Whatever the mechanism of sleep may be, it goes far back in vertebrate history and seems to have its counterpart even in some invertebrates.

A blow on the head may cause loss of consciousness, possibly by setting off a chaotic mass discharge of nerve cells which itself has no pattern and momentarily dispels all other patterns. This is possibly also the explanation of loss of consciousness during the epileptic fit—a massive, unpatterned discharge of neural activity from some hyperexcitable focus. Lack of oxygen causes unconsciousness, because nerve cells cannot function without oxygen, and fainting is attributable in most instances to transient circulatory failure with a resulting drop in blood pressure and a sharp decrease in blood flow and oxygen supply to the brain.

Events reach the level of consciousness only *after* they have occurred, *never* simultaneously. Sensory nerve impulses must travel from the periphery to the central nervous system, and be distributed to one or more sensory areas in the cerebral cortex, before they can enter the field of consciousness. These impulses travel in the human body at a rate of 3 to 30 meters per second (6 to 60 miles per hour), and they are further delayed at every synaptic junction by some three thousandths of a second, so that the time elapsing between the application of a stimulus (say an electric shock lasting only one thousandth of a second) and its recognition in awareness ranges from one-tenth of a second to more than one second. Even consciousness of a deliberate movement of the hand or foot is equally removed in time from the motion itself. The movement follows only after the interval required for the transmission of nerve impulses from the motor cortex to the periphery, and since what is customarily perceived is the contraction of muscles or change of position of the hand or foot, the sensory impulses recording the movement are further delayed by transmission from the periphery back to the cortex.

Both sensory and motor sequences are, however, given the semblance of simultaneity by their fusion and persistence in areas of the brain that are several steps removed in neurologic level (and in time) from the initiating events themselves. It is this fusion and persistence that give to consciousness its seeming continuity from one moment to the next.

Those things that dominate in consciousness are commonly said to occupy the 'center of attention,' but even here the clarity and intensity of awareness vary from moment to moment. Around the focus of attention is a peripheral field functionally belonging to consciousness, but not to immediate awareness; in this shadowy area events may leave their mark in unconscious memory, to be recalled only under hypnosis or in association with special circumstances. This dim penumbra is the area that the psychiatrist calls the 'unconscious [or subconscious] mind.' These are meaningless words physiologically. We repeat Hume's definition of 'mind' as an abstract term denoting the series of ideas, memories, and feelings which appear in consciousness: it is meaningless to think of an abstract term as being either 'unconscious' or as 'subconscious.' These words represent a mistranslation into English of Freud's German *unbewusst*, which literally means 'unbeknown.' Freud meant that memories may be primarily established or secondarily repressed away from the focus of immediate awareness—away, so to speak, from the center of the screen. His great contribution to the science of man was his demonstration that under certain circumstances these 'unbeknown' memories resist recall, or are incapable of recall into awareness, and yet they can modify conscious, ideational, and emotional life and thus determine many behavior patterns. But the 'unbeknown,' like the known, is also a matter of degree, of more or less easy recall into the focus of attention.

A man is momentarily, however belatedly, conscious of his immediate voluntary actions, but the conscious record is sometimes so faint and transient that it is as readily forgotten as a dream and he finds himself behaving like an automaton. He may spend a busy day at the office engaged in seemingly intelligent conversation with his secretary and a dozen associates, answer the telephone, read his mail and write letters, and then drive five miles through congested traffic and stop for innumerable red lights on his way to dinner, without remembering one-tenth of his voluntary and (at the time) fully conscious actions; and by the time he has had dinner he may have completely forgotten where he parked his car, if indeed he can recall whether he drove the car uptown. These consciously directed but unconsciously mediated actions are called 'automatisms.' (Such automatisms must be clearly distinguished from instincts, which are genetically predetermined, specific, stereotyped, and invariable patterns of behavior characteristic of a given species, the acquisition of which does not depend on learning, as does the acquisition of all automatisms. Man has relatively feeble instincts as compared with the lower animals, and these are dominant in his behavior only in the first months and years of life.)

In the most complicated patterns of voluntary activity consciousness may play a very limited role. A good example for discussion is the piano, because the most intricately and perfectly co-ordinated of all voluntary movements in the animal kingdom are those of the human hand and fingers, and perhaps in no other human activity do memory, complex integration, and muscular co-ordination surpass the achievements of the skilled pianist. In the early stages of learning, the novice strikes each note only after conscious deliberation: In effect he says to himself, "Now I will strike this note, now I will strike that one—" until, as he acquires competence, he finds himself anticipating the future: "After I strike this note then I will strike that one . . ." and so on until he gains command both of his fingers and the score and the conscious contribution can be reduced to, "Now I will play this movement generally *piano* and *largo* because I think that is the way that the composer meant it to be played." A musician who has to concentrate on the mechanical details of his music is not expert. With competence he gives thought only to his interpretation. At the orchestral cue, his fingers spell out the harmony and rhythm so faultlessly that he has scarcely a sense of playing the smashing chords and complicated arpeggios, remembering each note for only a fraction of a second (though let him strike a wrong one and he will remember it!) as he gives attention to the smooth interchange of melody between the orchestra and his instrument, and to the thousand and one matters affecting interpretation. In retrospect he is perhaps more keenly aware of the quality of his performance, which is just as complicated a matter as the mechanics of the performance itself, than of the innumerable mechanical details.

Sir James Paget, a noted nineteenth-century British surgeon, once timed a friend, Mlle. Janotha, while she played a *presto* by Mendelssohn, 'one of the fastest pieces of music known to her.' (This was presumably

the 'Perpetuum mobile,' Op. 119.) She played 5995
notes in four minutes and three seconds, or more than
twenty-four notes per second. Recognizing that each
note required at least two voluntary movements—flexion
and extension—as well as lateral movements in either direction, Sir James estimated that no less than 72 distinct
motor actions were required per second, each accurately
timed and exercised with judgment. (This calculation
takes no account of the movements of the hand, forearm, arm, and foot.)

The Mendelssohn *presto* does not, however, present
the most complex kinds of difficulties that are to be encountered in piano music, difficulties that separately and
collectively serve to retard the over-all speed set by the
limitations of either the keyboard or the fingers. As
demonstrated to the writer by Dr. David Saperton, a
five-note 'blind trill,' or tremolo, with each hand playing alternate notes, can be played at a rate close to 80
notes per second. This is not necessarily the greatest
speed, though it is one compatible with good musical
rendition. At this level, speed may be limited by the action of the piano, since the resistance offered by the key,
the depth to which the key must be depressed, the
length of the lever from tip to fulcrum and fulcrum to
hammer, the speed of rebound of key and hammer, and
other mechanical features, all contribute to the limitations imposed by the keyboard itself. A greater speed
can possibly be achieved with certain types of electronic
organs where mere contact between the finger and the
key may suffice to produce a tone, and it is conceivable
that, per finger, rapid chromatic passages on the violin
may exceed in speed the fastest possible action on the
piano.

A rapidly executed trill is, however, only one of the
innumerable movements that a musician must execute,
which include the wide spacing of keys to be struck in
succession by a single finger, and hence wide lateral displacements of both the finger and the hand or arm, as

well as forward or backward displacements of the finger or hand in passing from the white to the black keys and back. Each of these motions requires time and offers an additional impediment to speed; and consequently the introduction of harmony and melody, the compounding of thematic passages, of two or more voices, and of connected phrases, all serve to slow the music.

Fingering reaches what is perhaps its greatest complexity in the works of Leopold Godowsky (1870–1938). For example, in this composer's 'Badinage,' Chopin's 'Black Key' Etude (Op. 10, No. 5) is alternated with his 'Butterfly' Etude (Op. 25, No. 9) in the right and left hand, respectively, with preservation of the original melodies; both hands are working furiously throughout and operating against all the resistance factors enumerated above. As recorded by Dr. Saperton for Command Performance Records (No. 1202), this work, containing 1680 independent finger movements, requires 80 seconds, or involves an average of 21 notes per second. The execution of this work is not made easier by the constant use of two notes against three (polyrhythms), sometimes with two in the left hand and three in the right, and sometimes the reverse. There is no doubt that Godowsky's 'Badinage' is far more difficult to play than the Mendelssohn *presto*, as is attested by the fact that the former, like much of Godowsky's music, is rarely heard in public performance, and to the writer's knowledge his most difficult compositions have not been recorded by anyone except Dr. Saperton.

In this same category is Godowsky's arrangement of Strauss's 'Artists' Life' waltz which, in the original, has a different theme for each movement. Godowsky often fuses two or three of these themes into a simultaneous pattern while maintaining a three-four rhythmic motif in the bass, and when in one place a vertical or harmonic factor is added, there are five factors in simultaneous execution, played with such individual emphasis that the separate themes can be distinguished. Here the achieve-

ment not only involves finger speed and complex motions of the hands, but the control of accent on individual notes interspersed throughout the generally lively
pattern.

Chopin is well known for his complicated passages,
among which musicians frequently mention the *presto*
from the B flat minor Sonata. This passage has 1760
notes, and, as played for Victor Recordings by either
Sergei Rachmaninoff or Artur Rubinstein, requires 1 minute and 16 seconds. This is a speed of 23 notes per second. Chopin's Etude Op. 25, No. 11 in A minor (the
'Winter Wind' Etude), exclusive of the first four slow
measures, has 2043 notes in the right hand and 1081 in
the left. In Dr. Saperton's recording (Command Performance Record No. 1203), this work requires 3 minutes and 1 second, a speed of 11 notes per second for
one hand, and 16 notes per second for both—the speed
is here limited by wide excursions of the hand and a
relatively slow bass, but the over-all pattern is one of
great complexity.

Schumann's C major Toccata, Op. 7 (Victor Record
No. 14263) has 6266 notes and, as played by Simon
Barère without the repeat, requires 4 minutes and 20
seconds—a speed of 24.1 notes per second. The third
movement of Weber's Sonata No. 1 in C major, Op. 24,
contains 4700 notes and was played for the writer by
Mr. Randol Masters in 3 minutes and 50 seconds, or at
a rate of 20.4 notes per second.

In such works as are cited above, the impediments to
speed are great, and as Dr. Saperton demonstrated for
the writer, there are several fairly complicated passages
of piano music that can be executed at 29 to 33 notes
per second, though this speed is not compatible with
their good musical interpretation.

The examples cited show, however, that the upper
range of fingering is of the order of 20 to 30 notes per
second. We believe that Sir James Paget, in assigning
three muscle movements to each note, underestimated

the complexity of the problem. Practically all movements of a finger involve all three joints, so that a single note involves at least three motor nerve volleys for flexion, three for extension and at least one for lateral movement. Moreover a finger that is motionless is not in a state of inactivity but is tensed into position by the opposition of flexor and extensor muscles, and for any finger to go into action, at least two fingers must be moved out of the road, involving another fourteen motor actions. So without counting the motions of the wrist, forearm, shoulder, and trunk, or those involved in the use of the pedals, a speed of 20 to 30 notes per second may involve 400 to 600 separate motor actions—all effected by a competent musician with such automatism that he can give his attention to the over-all effects, rather than to the mechanical details.

Despite the fact that so much of this complicated muscular activity is at the level of automatism, complete automatism constitutes an inferior performance in no way above the level of a mechanical piano. In this instrument the limited expressive devices do not include the many artifices which the artist uses with such variety. The artist achieves his interpretation by minute deviations from uniformity and regularity: one note is held longer than the metronome permits, another is hurried; one note is struck firmly, another only lightly touched; a succession of notes each having equal paper value may be played with uneven tonal emphasis, or slurred deliberately, or made to parade in a presumptuous or sensuous manner. Huneker, in his introduction to the Chopin mazurkas, tells a story about someone accusing Chopin of writing in 3/4 time and playing in 4/4 because he prolonged the second beat; Chopin merely shrugged his shoulders and replied that it was a national trait. The stressing of the second beat is a characteristic of the mazurka, a Polish national rhythm.

Conversely, in reducing a melody to paper, the theme, which may be nascently tenuous and undisciplined, must

be rigidly molded into measures, beats, and rests, and
made to ascend and descend on the limited ladder of
the conventional octave—interpretation is, in part, a matter of liberating Orpheus's inspiration from this regimen-
tation between vertical and horizontal lines.

The 'meaning' of music is carried by a variety of elements: melody, rhythm, pitch, timbre, sequence, harmony, anticipation, suspension and fulfillment. Its essence is tonal form, and is not necessarily related to our
experience of words, things, or events. Much good music
is programmatic, but music need not be programmatic
to be good. Nor is good music necessarily descriptive of
any one of the emotions, or intended to evoke any particular emotional response: on the contrary, some critics
hold that music is at its best when, in composer, artist,
and auditor alike, it has been filtered by psychic distance and detachment until its affective qualities are
uniquely its own. Then it is that it conveys—or perhaps
elicits is a better word, because it is hard to say what a
composer intends to convey—its own semantics, its own
affirmations and denials, syntheses and antitheses, in
pilasters and arabesques of sound, substituting myth for
reality, complementing, not imitating, other forms of
perceptual experience with values of its own.

But these musical values are not absolute, in the sense
of having the same significance for all individuals—as is
witnessed by the changes in musical style with the passage of time, and the wide and sometimes vehement
division of opinion with respect to style and content.
These values, both in respect to creation and reception,
vary in different individuals with their emotional
make-up, their musical experience and ability, their
prevalent interest and emotional set, their intellectual
maturity, and perhaps with such fundamental psycho-
logical matters as how they hear music—as a plastic flow
of sound, as an increase or decrease of intensity, or as
isolated notes like blocks of stone placed at different ele-
vations. It is the ever-conscious task of the artist to pre-

serve such values from being lost in automatism, and to infuse new values stemming from his own interpretation.

Thus, Dr. Saperton cogently emphasizes the 'ratio' between conscious attention and mechanical performance:

"I believe," he says, "that there is a constantly varying or shifting ratio between the 'automatic' and the 'aware.' This ratio differs with different individuals and at different times within the same individual. But a musically fine performance in the highest re-creative meaning is *directed*, in the artistically critical sense, by the controls or factors within the area of 'directed awareness.' I do not believe that a performance of high artistic quality can be produced at the level of pure automatism.

"Which means, of course, that since two matters cannot occupy the center of attention simultaneously, the attention of the artist must oscillate with remarkable rapidity from one matter to another throughout his performance."

At the other extreme of mental agility are those occasional individuals who perform the most complicated mental acts with no special training and no conscious deliberation whatever—a recent example to reach public attention being Shakuntala Divi, the Brahman girl who quickly extracted the 20th root of a 42-digit number, or multiplied figures yielding 39 digits, without hesitation and with no knowledge of how she did it.

Consciousness is not and never has been a prerequisite for function in the nervous system; it only supplements the activities of the nerve net, the nerve cord, the visceral nervous system, the brain stem and cerebrospinal reflexes, all of which can and do function, even in man, at the unconscious level. Consciousness is a very special, and very specialized, function of the nervous system. Its unique value lies in its delay time, its sluggishness, its time-binding quality: it carries a residue of neural activity from one instant to the next, giving the semblance of continuity to what are, in actual fact, extremely brief and isolated events in the nervous system. In this sense it may be likened to a television tube which glows for a fraction of a second after it has been electrically excited, and thus affords a continuous instead of a flickering image.

The psychologist William James was one of the first to espouse the 'natural history' of consciousness and it was also he who (after Hume) introduced the 'stream of consciousness' motif into psychology and literature. But the kaleidoscopic fragments presented in the stream of consciousness comprise only one-half of the problem —What binds these fragments into a lesser or greater whole, into a temporal continuum? To return to Hume's point, in the appendix to his *Treatise of Human Nature* (1739) he suddenly realized that in confining all cognition to single perceptions and supplying no faculty for unifying, recording or classifying these perceptions, he had destroyed the possibility of knowledge:

'When I turn my reflection on myself, I never can perceive this self without some one or more perceptions; nor can I ever perceive any thing but the perceptions. 'Tis the composition of these, therefore, which forms the self. . . . But all my hopes vanish, when I come to explain the principles, that unite our successive perceptions in our thought or consciousness. . . . In short, there are two principles, which I cannot render consistent; nor is it in my power to renounce either of them, viz. that all our distinct perceptions are distinct existences, and that the mind never perceives any real connexion among distinct existences. Did our perceptions either inhere in something simple or individual, or did the mind perceive some real connexion among them, there would be difficulty in the case.'

Today neurophysiologists, generally rejecting, as Hume did, anything more than a figurative meaning for the word mind, are scarcely any closer to resolving the difficulty presented by temporal continuity in consciousness than was the philosopher. They possess, however, the vantage of the evolutionary point, not available in the eighteenth century, and are prepared to look upon consciousness (as man knows it subjectively) as the very complicated product of a very long process of evolution.

As the human understanding surpasses that of the ape, and that of the ape surpasses that of the fishes, so in almost as extreme a degree the vertebrate brain surpasses the nervous organs of the invertebrates. A reason for this difference, one thinks, is to be found in history, and in a very elementary biological fact. All animals are dependent on either plants or animals for food, and from its beginnings the evolution of the animal kingdom has in the main presented a pageant of predator and prey—eat or be eaten! In the Cambrian period, which opened some 550 million years ago, there may have been many soft-bodied animals which are not preserved in the fossil record, but most of those which are preserved possessed destructive mouth parts, prehensile limbs and muscular appendages by which they could pursue their prey, and they were themselves protected from their enemies by chitinous or calcareous exoskeletons. The mobile, predatory habit (with obvious exceptions for scavengering, vegetarianism and parasitism) early became what we may call the typical habitus of the animal kingdom, and this habitus required that the successful animal solve the problem of two moving bodies—its own and that of its prey—and not just in the three dimensions of space, but in a fourth: accurate timing was a *sine qua non*, and accurate timing required the integration of events of the recent past with those of the present moment, thus permitting extrapolation into the future and assuring both food and safety.

If we set to one side, as supported by insufficient evidence, all such means of foreseeing the future as crystal balls, palm reading, lot casting, astrology, hepatoscopy, table tipping, and clairvoyance, the future can be anticipated only by the permutation and recombination of recollected past experiences. This is a game that consciousness plays—or that plays itself in consciousness—and that can be called 'take and put':—if I take this and put it there, then so-and-so may happen. Man plays this game in its most elaborate form with abstractions, symbols, checkers, chessmen, letters, words, logical and mathematical relations, hypotheses; in this sense, namely in the use of abstraction and symbolization, the chimpanzee is a novitiate, the racoon, dog, cat, and rat in diminishing degree show nascent talent, and below the mammals the game in this form seems scarcely to be played at all: to play it requires, first of all, recognition of the fact that the game exists, that one *can* 'take and put,' plus the capacity to form abstract concepts. But the game of 'take and put' does not basically involve symbols or abstractions—it has its roots in a simple awareness which is charged with plus- and minus-values, and involves at most the simple motor action of going from here to there. The white rat does not reason, but it can solve problems involving 'triangularity' and 'circularity' by the trial-and-error method, and can remember the solution—if the solution is important to its well-being. At an even lower level, the bird can make an accurate two-point landing on a moving limb, because it is important for it to do so; the fish can strike its swiftly moving prey because it is hungry, and can judge distance, direction, and velocity This is the game of 'take and put,' of going from here to there, in its elementary form.

The heavily armored, sluggish ostracoderms were certainly aware, in a dim sort of way, of where they were going, and why, as they squirmed over the muddy bottoms of the Ordovician-Silurian lakes and rivers. Even at this level we can speak of 'voluntary action,' in so far as the word 'voluntary' indicates some ability to choose between conflicting impulses and to assess the consequences of possible action in terms of past experience and anticipated future. With the emergence of the placoderm fishes, going from here to there became both easier and more complicated, because spines had given way to fins, and the mud bottom had given way to a world of up-and-down and left-and-right. As the round mouth of the ostracoderms was replaced by jaws, capable of catching and crushing other fast-moving animals, life in this three-dimensional world fostered increased speed, accuracy, and cunning—a moving target requires prejudgment of the future, as well as perception of the past and present. When the Amphibia crawled out of the water onto the banks of the Mississippian lakes and rivers, the distances between one clump of rushes and another, the feel and smell of moist or dry ground, required new elements of judgment, as did the trick of wrapping a tongue around a careless dragonfly. The reptiles, when they left the watercourses to lay their eggs on land and to grub on land for food, had to assess lights and shadows, colors, shapes, humidities and temperatures, all with some skill, and they had better sense-organs for these purposes. The homeothermic mammals had to look out not only for themselves but also for their young: the nest, the lair, the hideaway, became a nucleus around which sights and sounds and smells wove a tangled web challenging sensory acuity, quick integration, accurate anticipation. At each stage, natural selection favored the animals that had an increased capacity to see their way from here to there, and to anticipate the consequences of going from here to there.

This view affords us a somewhat better definition of consciousness, which we can now designate as awareness of environment and of self, revealed objectively by self-serving, neuromuscular activity which exhibits choice between alternative actions and simultaneously relates past experience to anticipated future. Without temporal persistence in perception—without what we have called the 'time-binding' quality of consciousness—

to tie one moment to the next and to afford a basis
for anticipating the future—all neuromuscular reactions
would be essentially instantaneous and of limited value
for survival of the organism. Whether the time-binding
activity extends over a period of seconds or of years is
immaterial to the cogency of the definition.

The evolution of the vertebrates thus leads us to an
unavoidable conclusion: consciousness is not unique to
man, to the primates, or to the mammals; it goes back
to the roots of vertebrate history and has been progres-
sively elaborated in content, coloring, and complexity
roughly in proportion to the evolution of the neuromuscular system. It cannot even be argued that consciousness is a unique vertebrate invention—the crab, the octopus, the butterfly, the ant, the bee all possess sensory
devices imparting to them the awareness of the world,
and all demonstrably engage in integrated, time-binding,
self-serving action; and it must be presumed that all participate in some proportional measure in conscious appreciation of themselves and their environment.

But accepting this definition, it can tentatively be argued that the capacity for consciousness requires the
anatomical substratum of a nervous system of some sort,
with appropriate sensory, integrative, and motor com-
ponents, and one need not look for consciousness or any
simulacrum thereof in lower forms of life such as the
Protozoa even though they swim or crawl and show
simple avoidance reactions, until it is demonstrated that
they show some capacity for time-binding and the integration of diverse stimuli—which, by present evidences,
they do not.

On the other hand, it must not be supposed that because the mechanisms for integrated behavior are present, the animal is necessarily a conscious creature. On
the contrary, as is well demonstrated in man, consciousness requires a special type of neural activity. It can be
turned on and off, and this fact alone leads us to suspect
that sleep is a neural mechanism which has biological

survival value. The mechanism of sleep has been evolved, like all else, and is most highly developed where consciousness is most highly developed—in the mammals; but there is no reason to believe that consciousness is without its occasional blackout in submammalian forms.

Nor does it follow that in its lower levels of development consciousness assumes the elaborately integrated, egocentric human pattern known as 'self.' In man the 'self,' the seemingly enduring spectator-director who commands the performance, is an impermanently sustained pattern of neural activity. Far from an entity enduring from day to day, it is a flickering image formed where the rays of sense are brought to focus in the conscious pattern; it forms and dissolves in successive instants, and never re-forms the same. Pathologically it can suffer such distortion or dissociation of its parts as to lead to the tragic fractionation of personality.

Man's body, as Sherrington pointed out, has genetic continuity through his chromosomes with his predecessors throughout the geologic ages, but there is no such genetic continuity in consciousness. There is no requirement for either self-awareness or awareness of environment in the individual organism until it develops to the point where it possesses both physical freedom and the neuromuscular system necessary for it to take advantage of that freedom; and there is no evidence of consciousness in the ovum and sperm, or in the immature embryos of the fishes, Amphibia, reptiles or birds, where it would be of no avail; nor is there any evidence of its presence in the entire nonmotorized plant kingdom. It is plausible to believe that self-awareness begins to stir in the mammalian fetus at about the time when the developing organism begins to kick and squirm, simply because this is the time of beginning motor activity when, even *in utero*, something is to be gained by going from here to there.

Many writers localize consciousness in the cerebral cortex; but this localization must be qualified because sub-mammalian forms that have no cortex present all the evidences of conscious awareness of themselves and of their environment, and even in man it can be demonstrated that the brain stem contributes to conscious states. It is probably more accurate to say that consciousness is a function of the highest levels of the central nervous system, but invariably includes contributions from the brain stem and even from the peripheral nervous system. Nevertheless, there is every reason to believe that consciousness in man reaches its highest integration in the cortex, and neurophysiologists have long been concerned with the question as to whether any particular region of the cortex is important for conscious activity.

The cortex is divided longitudinally into two halves, corresponding to the right and left cerebral hemispheres Most of the sensory pathways entering the cortex, as well as the motor pathways leaving it, cross the central nervous system at some point—so that, functionally, the left hemisphere is predominantly related to the right side of the body; the right hemisphere, to the left side Although there is only a slight difference in the size of the cerebral hemispheres, the left hemisphere dominates cerebral activity in a right-handed person; the right hemisphere, in a left-handed person. This dominance is particularly evident in respect to manual skill, symbolic understanding, and language. Dominance is relative, however, rather than absolute, and surgical denervation, or removal of one cerebral hemisphere, may be followed, after prolonged training, by the development of increased function on the other side.

Each hemisphere of the cortex is somewhat arbitrarily divided into four lobes: the frontal, temporal, parietal, and occipital. This division is mostly a matter of convenience in description because all parts of the cortex are directly or indirectly connected with all other parts by association fibers, and no region can be regarded as the exclusive seat of any particular cortical function. In a qualified sense, however, there is a greater degree of cortical localization in man than in the ape, and more in the ape than in the white rat. The neurosurgeon has thoroughly explored the cortex by means of electrical stimulation or other methods during operations on the brain, and has thus identified numerous areas which are more or less specifically involved either in sensory or motor activity, and which are functionally related to specific parts of the body. This localization is, however, in great part a matter of functional pattern rather than of anatomy, as is shown by the extent to which it changes during the growth of the individual. If one were to outline on the head of a very young infant the areas concerned with each subdivision of the body, the resulting homunculus would consist of a great suctorial mouth and tongue and a big nose almost entirely surrounded by the two hands—because these are the parts of the body with which it is importantly concerned throughout its early life. Only with the passage of some months do the hand and mouth areas shrink as the eyes, ears, and feet come to occupy attention, and only later do the shoulders, toes, arms, and thighs, and lastly the back, acquire significance. Furthermore, the very young infant qualitatively resembles the ape and rat, in that localization is not very important for function, and widespread substitution is possible. But with the growth of the brain, which increases in size fourfold between birth and adulthood, localization becomes more marked and increasingly important, so that in an older adult the destruction of a small area in one hemisphere may have severe and permanent consequences for function on the opposite side of the body.

Of all the areas in the cortex that have puzzled neurophysiologists and neurosurgeons, the most intriguing are the so-called 'silent areas,' which, on stimulation, evoke neither sensory nor motor response. The frontal lobes in man consist, in large part, of such silent areas. It is in

the relatively great increase in mass of these frontal lobes
that man's brain differs from the brain of the apes, and
it is the great development of the frontal lobes that
replaces the receding forehead of the ape by the pro-
truding forehead in man.

Uniquely developed as they are in man, the frontal
lobes present a mystery. Not only are they inexcitable
electrically, but the nerve tracts leading to them can be
cut (leucotomy) with relatively slight effect on intelligence. Leucotomy was first performed on man to give
relief from intractable pain, and then it was discovered
that the operation could ameliorate certain forms of
mental illness, and for this latter purpose the operation
has now been performed on many thousands of patients.
In respect to intractable pain, the pain remains after the
operation, but it no longer causes any distress; it no
longer hurts, it is just there; the patient perceives his
pain with as much indifference as if he were perceiving
something outside his body. In respect to mental illness,
psychiatrists agree that agitation, depression and obses-
sion may sometimes be relieved, and by so much the
patient may be restored toward normal. But otherwise
the only notable effects are a reduction of restraint, judgment, initiative, tact, and foresight; the patient tends to
become lazy, fat, carefree, a little silly, and to lose his
inhibitions. Thus the frontal lobes appear, first, to im-
part a distressful quality to pain; and second, to project
a man beyond his immediate circumstances in space and
time, and to give him drive, and ethical and other types
of valuation. Perhaps these are the features that chiefly
distinguish human consciousness from that of the great
apes.

The view presented here, that consciousness is a specific
concomitant of the evolution of motor activity, finds support in recent trends in neurology. The neurophysiologist, R. W. Sperry, in analyzing the neuroanatomic basis
of conscious experience, emphasizes that the actual neu-

ral patterns set up in the brain cannot correspond, by any sort of point-to-point correspondence, to either the external objects of sense or the subjective figures of consciousness. Nor is localization of neural activity in particular parts of the brain the secret of specific sensory qualities: sensations as diverse as those of red, black, green, and white—or, alternatively, of touch, cold, warmth, movement, pain, posture, and pressure—may arise during activation of the same cortical areas—the visual areas, on the one hand, and those involved in muscle sense, on the other. The explanation of sensory form and quality must rather be sought in patterns of activity involving large areas of the brain as a whole.

When one perceives an object, one is thereby prepared to respond with reference to it in some manner or other—by pointing to it, by outlining it with a finger, by relating its location in space relative to other objects, by liking or disliking it; or, at the level of the white rat, by avoiding or approaching it, by running under it or up one edge, by leaping to a corner of it, by smelling or biting it. 'Perceiving' and 'being-set-to-act' (or refusing to act) are to be equated with each other. It does not matter into what sensory areas of the cortex the perceptive pattern drifts; the preparation-to-respond remains the same, and hence the meaning, the value, remains the same.

The trained student frequently, the layman generally, assumes that the major function of the brain is to manufacture ideas, sensations, images, feelings, to store memories, and the like. On the contrary, Sperry sees these phenomena as by-products: the immediate purpose of brain function is to transform sensory patterns into patterns of motor activity, and on close analysis the sensory pattern proves basically to be but a means to the end of motor co-ordination, in that it allows some additional refinement, such as the appreciation of past experiences or of future goals, to be brought into behavior, thereby increasing over-all adaptiveness.

Since the publication of the *Origin of Species* the ever-increasing evidences afforded by the sciences of life have only served to emphasize the unitarianism of nature—the fact that the cosmos is a 'universe' and not a 'diverse.' Many years ago the British physicist, John Tyndall, speaking of matter *versus* life, said in his famous Belfast Address (1874), 'Let us reverently, but honestly, look the question in the face. Divorced from matter, where is life? Whatever our faith may say, our knowledge shows them to be indissolubly joined.' We can paraphrase him and ask, 'Divorced from matter, where is consciousness? Whatever our faith may say, our knowledge shows them to be indissolubly joined.' Tyndall, only fifteen years after the publication of the *Origin*, saw that 'The doctrine of evolution derives man, in his totality, from the interaction of organism and environment through countless ages past. The Human Understanding . . . is itself a result of the play between organism and environment through cosmic ranges of time.'

The intrinsic responsivity of protoplasm and the primitive nerve net could have evolved in many other directions of 'awareness,' as indeed it did in the starfish, clam, worm, crab, lobster, scorpion, centipede, spider, butterfly, ant, and bee. But in the vertebrates it happened that evolution followed another course because their ancestors radiated from the sea into the unstable waters of the Cambrian-Ordovician continents; becoming encased in armor, they evolved jaws and fins and legs, and a kidney that afforded them a stable internal environment in which to live and in which the brain could function at the highest integrative level. Psyche and soma have come up the long road of vertebrate evolution together, seeking the free and independent life, seeking always to minimize life's disquietudes.

All that remains in man of the ancient armor of the ostracoderms is represented by his teeth, nails, hair, and skin. However, through the ostracoderms and the vertebrates that followed them, he has come to add conscious purposes and plans to his devices for minimizing his disquietudes. And thus he has come to say, 'I am . . . I feel . . . I know,' and to ask himself:

'How is the MEness, this consciousness that peers through a window past a hedge of *Rosa rugosa* and a lawn bounded by a stone wall, to the blue waters of Frenchman Bay and a clouded sky, related to the atoms and molecules that comprise my brain, and specifically to those areas in my cerebral cortex where I do most of my thinking?'

It scarcely suffices to say that a capacity for consciousness is inherent in some primordial form in every atom and molecule, because, of a total of 92 elements in the periodic table (not all of which have been isolated)— and not including ten transuranyl elements (neptunium, plutonium, americium, curium, californium, berkelium, einsteinium, fermium, mendelevium, and one tentatively named nobelium) which have been artificially knocked together in a cyclotron—probably not more than 12 enter into the composition of my brain: carbon, hydrogen, oxygen, nitrogen, phosphorus, sulfur, sodium, potassium, calcium, chlorine, iron, and magnesium. Nor does it help to equate 'mind' and 'matter' at the level of the electron —as, in effect, A. N. Whitehead does; why choose the electron, when there are at least 30 other 'fundamental particles' to choose from?

Moreover, it is clear that my consciousness resides not in any particular atoms because the atoms that are part of me today, tomorrow will be gone from me to be replaced by others—there is scarcely one that is 'mine' for more than a few weeks at most—but rather in their unique and transient patterns of activity.

The bald facts are that, for matter to know itself in ME, ten billion neurons in my brain, and many, many times that number of functional connections, are required to give me the past, the present, and the all too inaccurately divined future that contrive this moment.

Five hundred million years of vertebrate evolution have been required to produce this brain, composed of a dozen-odd sorts of atoms; and, given an adequate internal environment, it can know itself in self-awareness for at most some three- or four-score years.

How all this comes about in the transient interplay of atoms and molecules, no one can answer, but this does not mean that the answer is unattainable. Science advances so rapidly that it would be rash to place a limit on its possibilities. In essence, the scientific method consists of careful observation of nature and cautious confirmation of all conclusions, to the exclusion of unsubstantiated hypotheses. A scientist is one who, when he does not know the answer, is rigorously disciplined to speak up and say so unashamedly; which is the essential feature by which modern science is distinguished from primitive superstition, which knew all the answers except how to say, 'I do not know.' On every scientist's desk there is a drawer labeled UNKNOWN in which he files what are at the moment unsolved questions, lest through guesswork or impatient speculation he come upon incorrect answers that will do him more harm than good. Man's worst fault is opening the drawer too soon. His task is not to discover final answers but to win the best partial answers that he can, from which others may move confidently against the unknown, to win better ones.

The atomistic basis of consciousness remains unknown. But by all the evidences it must be sought at the cellular rather than the molecular level.

There are those who say that the human kidney was created to keep the blood pure, or more precisely, to keep our internal environment in an ideal balanced state. This I must deny. I grant that the human kidney is a marvelous organ, but I cannot grant that it was purposefully designed to excrete urine or to regulate the composition of the blood or to subserve the physiological welfare of *Homo sapiens* in any sense. Rather I contend that the human kidney manufactures the kind of urine that it does, and it maintains the blood in the composition which that fluid has, because this kidney has a certain functional architecture; and it owes that architecture not to design or foresight or to any plan, but to the fact that the earth is an unstable sphere with a fragile crust, to the geologic revolutions that for six hundred million years have raised and lowered continents and seas, to the predaceous enemies, and heat and cold, and storms and droughts; to the unending succession of vicissitudes that have driven the mutant vertebrates from sea into fresh water, into desiccated swamps, out upon the dry land, from one habitation to another, perpetually in search of the free and independent life, perpetually failing, for one reason or another, to find it.

There are those who say that they can extrapolate from purpose in the organism to purpose in the cosmos, from personality in man to a personality transcending the stars and nebulae. This, I must question. Purpose in the organism issues from its molecular structure, as does personality in man; and both are transient patterns in the swirling fountain of matter and energy that in a few thousand million years has spewed galaxies in inconceivable numbers and at inconceivable speeds into the impenetrable depths of space. It is scant modesty for man, even if he is the 'highest vertebrate,' to presume that he can predicate the cosmic plan on the intensity of his joy or pain, or cement the stars together with even his highest aspirations.

No less than his lowly predecessors, he continues in the search of the free and independent life, for it is in the nature of all life to move into that equilibrium where the totality of desire is balanced against the totality of restraint. But because he is the highest vertebrate he can do what no other vertebrate can do: when, out of whatever desire and knowledge may be his, he makes a choice, he can say 'I will . . .'

And knowing how and why he says 'I will' he comes into his own as a philosopher.

TECHNICAL NOTES AND
BIBLIOGRAPHY

The following notes supplement, for the technically interested reader, certain general statements in the text. Bibliographic references are limited to recent sources and to those specifically used in the text and notes.

I. EARTH

1. BERNARD, C. *Leçons sur les Propriétés Physiologiques et les Altérations Pathologiques des Liquides de l'Organisme.* Baillière, Paris, 1859, 2 volumes.
2. BERNARD, C. *Leçons sur les Phénomènes de la Vie Communs aux Animaux et aux Végétaux.* Baillière, Paris, 1878–1879, 2 volumes.
3. BROOKS, C. E. P. *Climate through the Ages.* Ernest Benn Ltd., London, revised edition, 1949.
4. BROWN, H. The age of the solar system. *Scientific American,* 196: (4) 80. 1957.
5. CANNON, W. B. *The Wisdom of the Body.* W. W. Norton & Company, Inc., New York, 1932.
6. EARDLEY, A. J. The cause of mountain building—an enigma. *American Scientist,* 45: 189. 1957.
7. EMILIANI, C. Ancient temperatures. *Scientific American,* 198: (2) 54. 1958.
8. GAMOW, G. *Biography of the Earth, Its Past, Present and Future.* Viking Press, New York, 1948.

9. HOLMES, A. A revised estimate of the age of the earth. *Nature*, 159: 127. 1947.

10. HOLMES, A. The age of the earth. *Endeavour 6:* 99. 1947.

11. KNOPF, A. Measuring geologic time. *Scientific Monthly*, 85: 225. 1957.

12. LINNENBOM, V. J. Radioactivity and the age of the earth. *Journal of Chemical Education*, 32: 58. 1955.

13. OLMSTED, J. M. D., AND E. H. OLMSTED. *Claude Bernard and the Experimental Method in Medicine.* Henry Schuman, New York, 1952.

14. PATTERSON, C., G. TILTON AND M. INGHRAM. Age of the earth. *Science*, 121: 69. 1955.

15. ROMER, A. S. *The Vertebrate Body.* W. B. Saunders Company, Philadelphia, 1949.

16. RUBEY, W. W. Geologic history of sea water. An attempt to state the problem. *Bulletin of the Geological Society of America*, 62: 1111. 1951.

17. SCHUCHERT, C., AND C. O. DUNBAR. *A Textbook of Geology. Pt. II. Historical Geology.* John Wiley & Sons, New York, 1941.

18. SMITH, H. W. The Evolution of the Kidney. In *Lectures on the Kidney*, University Extension Division, University of Kansas, Lawrence, Kansas, 1939, second edition, 1943.

19. UREY, H. C. *The Planets. Their Origin and Development.* Yale University Press, New Haven, 1952.

20. WILSON, J. T. Geophysics and continental growth. *American Scientist*, 47: 1. 1959.

Claude Bernard first used the word *milieu* to indicate the blood plasma as the environment in which the individual lives—as opposed to the external environment —in his lecture notes of 1857, and the concept of the constancy of this environment was developed at some length in his essays of 1859 {1}. He later expanded the idea in his *Phénomènes* {2}, using the terms *milieu in-*

térieur to designate the total circulating fluids of the body—that is, the plasma plus the interstitial fluid.

In the early decades of this century Bernard's thesis was extended in many directions by Walter B. Cannon, who summarized his own studies in his justly popular book, *The Wisdom of the Body* {5}. Cannon called the *milieu intérieur* the 'fluid matrix' of the body. It was he who coined the word 'homeostasis,' and who pointed out (*a*) that the existence of a homeostatic state is itself evidence that agencies are acting, or ready to act, to maintain this constancy; (*b*) that it remains constant because any tendency towards change is automatically met by increased effectiveness of the factor or factors which resist the change; and (*c*) that such regulatory factors are generally under automatic control.

On the structure of the earth the writer has followed Gamow {8}, on paleogeology, Schuchert and Dunbar {17}, and on the age of the earth, Holmes {9, 10}. The dating of geological periods is roughly that given by Romer {15}, which differs only slightly from the estimates given by other authorities (as, for example, {3, 4, 11, 12, 14}), no two of whom are in complete agreement.

Sir George Darwin's theory of the origin of the moon by fission of the earth is described by Gamow {8}. This theory was opposed on mathematical grounds in 1930 by Jeffries, and it has encountered new opposition in a recent dissertation by Urey {19}. Darwin's theory has, however, so many attractive points in its favor that it still has the support of many students of cosmogony. Many factors bearing on mountain building are reviewed by Eardley {6}.

II. EVOLUTION

21. JEPSEN, G. L. "Orthogenesis," and the fossil record. *Proceedings of the American Philosophical Society*, 93: 479. 1949.

22. JEPSEN, G. L., E. MAYR, AND G. G. SIMPSON. *Genetics, Paleontology, and Evolution.* Princeton University Press, Princeton, N. J., 1949.
23. SCHEINFELD, A. *The New You and Heredity.* J. B. Lippincott Co., Philadelphia and New York, 1950.
24. SIMPSON, G. G. The problem of plan and purpose in nature. *Scientific Monthly,* 64: 481. 1947.
25. SIMPSON, G. G. *The Meaning of Evolution.* Yale University Press, New Haven, Conn., 1949.

The calculation of chromosomal and genic permutations is that of T. Dobzhansky, cited by Scheinfeld {23}, whose popular book is recommended to anyone interested in human genetics.

I have drawn heavily on many of G. G. Simpson's works; the 'alphabet' is cited from {24}, and quotations are drawn from this article and from {25}.

The issues of 'directed' evolution, vitalism versus materialism, and related topics, are discussed by Jepsen {21}; by Colbert, in 'Progressive adaptations as seen in the fossil record' in {22}; by Wright in 'Adaptation and selection' in {22}, and by Simpson {24, 25}.

III. THE PROTOVERTEBRATE

26. BARRELL, J. Dominantly fluviatile origin under seasonal rainfall of the Old Red Sandstone. *Bulletin of the Geological Society of America,* 27: 345. 1916.
27. BARRELL, J. The influence of Silurian-Devonian climates on the rise of air-breathing vertebrates. *Bulletin of the Geological Society of America,* 27: 387. 1916.
28. BERRILL, N. J. *The Origin of Vertebrates.* Clarendon Press, Oxford, 1955.
29. BRYANT, W. L. A study of the oldest known vertebrates, Astraspis and Eriptychius. *Proceedings of the American Philosophical Society,* 76: 409. 1936.

30. CHAMBERLAIN, T. C. On the habitat of the early vertebrates. *Journal of the Geology Society*, 8: 300. 1900.

31. DENISON, R. H. A review of the habitat of the earliest vertebrates. *Fieldiana* (Geology), 2: No. 8, August 1956.

32. GREGORY, W. K. The roles of motile larvae and fixed adults in the origin of the vertebrates. *Quarterly Review of Biology*, 21: 348. 1946.

33. GREGORY, W. K. *Evolution Emerging*. Macmillan Company, New York, 1951, two volumes.

34. GREGORY, W. K., AND H. C. RAVEN. Studies on the origin and early evolution of paired fins and limbs. *Annals of the N. Y. Academy of Sciences*, 42: 275. 1941.

35. HENDERSON, L. J. *The Fitness of the Environment. An Inquiry into the Biological Significance of Matter*. Macmillan Company, New York, 1913.

36. MACFARLANE, J. M. *The Evolution and Distribution of Fishes*. Macmillan Company, New York, 1923.

37. MARSHALL, E. K., JR., AND H. W. SMITH. The glomerular development of the vertebrate kidney in relation to habitat. *Biological Bulletin*, 59: 135. 1930.

38. NEAL, H. V., AND H. W. RAND. *Comparative Anatomy*. P. Blakiston's Son & Co., Inc., Philadelphia, 1936.

39. ROBERTSON, J. D. The habitat of the early vertebrates. *Biological Reviews*, 32: 156. 1957.

40. ROMER, A. S. Eurypterid influence on vertebrate history. *Science*, 78: 114. 1933.

41. ROMER, A. S. *Vertebrate Paleontology*. Second edition. University of Chicago Press, 1945.

42. ROMER, A. S. The early evolution of fishes. *Quarterly Review of Biology*, 21: 33. 1946.

43. ROMER, A. S. Fish origins—fresh or salt water? *Papers in Marine Biology and Oceanography*, p. 261. Pergamon Press Ltd., London.

44. ROMER, A. S. *The Vertebrate Story*. University of Chicago Press, Chicago, 1959.

45. ROMER, A. S., AND B. H. GROVE. Environment of the early vertebrates. *American Midland Naturalist*, 16: 805. 1935.

46. SMITH, H. W. The absorption and excretion of water and salts by marine teleosts. *American Journal of Physiology*, 93: 480. 1930.

47. SMITH, H. W. Water regulation and its evolution in the fishes. *Quarterly Review of Biology*, 7: 1. 1932.

48. WHITE, E. I. *Jamoytius kerwoodi*, a new chordate from the Silurian of Lanarkshire. *Geological Magazine*, 83: 89. 1946.

Some years ago L. J. Henderson wrote a charming and commanding book called *The Fitness of the Environment* {35}, in which he pointed to the unique properties of carbon, hydrogen, oxygen, water, carbonic acid, and other biochemically important materials, as all contriving to make life possible. Where there had hitherto been much talk of the 'fitness of the organism' to the environment, Henderson sought to emphasize that the properties of the environment (at the atomic and molecular level) uniquely prepare it to sustain the processes of life. Henderson was neither a vitalist nor teleologist and indeed he declared himself to be a forthright mechanist: 'Given matter, energy, and the resulting necessity that life shall be a mechanism,' he said, 'the conclusion follows that the atmosphere of solid bodies (such as the earth) does actually provide the best of all possible environments for life.' In assessing Professor Henderson's arguments, we may note that life fits its environment and that its environment fits life, because life has been spun out of the very atoms and molecules which Henderson took as 'given'; but if one presupposes the slightest change in the properties of any of these atoms, then matter would be different, and so, consequently, would life. We can be sure that they would still both 'fit' each other, even as now, but whether for better or for worse from man's point of view is a subject that must be left to speculation.

Confusion with respect to the origin of the vertebrates has crystallized in the classification of the Vertebrata (over 25,000 species) as a subphylum of the phylum Chordata, the three other subphyla in this category being the numerically insignificant Hemichordata (perhaps 50 species), of which the worm *Balanoglossus* had held greatest attention, the Urochordata (about 1400 species), which includes the tunicates or ascidians, and the Cephalochordata (about 25 species), of which the lancelet, *Amphioxus*, is the biologists' ideal, which 'if it hadn't existed, would have had to be invented,' as Neal and Rand say {38, p. 664}. This imposing classification arises from certain parallels in embryology or adult anatomy, and notably from the presence of 'gill arches' at some stage of development, and of a stiffened rod or notochord extending down part or the whole of the animal's body, the last supplying the character from which the term 'chordate' is derived. None of these throws any light on the origin or structure of the protovertebrate. The erection of the phylum Chordata to include the balanoglossids, tunicates, and *Amphioxus* along with the vertebrates is a taxonomic tour de force which gains nothing, and loses much, and it would be better to relegate the first three to an appendix well removed from the Vertebrata and entitled 'Mysterious Creatures of Undetermined Affinities.' As Neal and Rand conclude, 'The Chordata clue seems to lead us into a blind alley out of which the most promising exit is the way back.' Those interested in the origin of the vertebrates will read Berrill's recent book {28} with interest.

Two of the most interesting papers in American paleontology—Chamberlain's and Barrell's—concern the freshwater origin of the vertebrates. Until 1900 it had been accepted that the seat of vertebrate evolution had been the sea, from which terrestrial forms had emerged across the tidal strand. Favoring this presumption was the circumstance that those Devonian beds in which the oldest fish fossils are found in greatest abundance were accepted to be marine in origin. However, geologists had come to realize that much of the Old Red Sandstone (or Devonian) of the British Islands was of fresh-water origin, and spoke of it as having been laid down in 'fresh-water lakes' or 'inland seas'—an equivocal description that failed to stimulate paleontological reinterpretation since it was conceived that the ancient fishes could have migrated transiently into brackish estuaries and even into fresh-water lakes.

A complete reinterpretation of this problem was first suggested by T. C. Chamberlain in 1900 {30}, who pointed out that the earliest known vertebrates were predominantly confined to sediments of established or suspected fresh-water origin, along with the remains of land plants and fresh-water molluscs, and strangely absent from the indubitably marine deposits. Noting that the typical vertebrate pattern is uniquely fitted to propel the animal against a moving stream, he proposed that the earliest vertebrates had been evolved in the fresh waters of the continents.

In the decade following Chamberlain's paper the fresh-water origin of the Old Red beds was increasingly accepted by geologists, and more attention was given to the conditions of their deposition. The biological aspects of this problem were reviewed in 1916 by Barrell {26, 27} who had himself contributed importantly to the geological interpretation of the Old Red deposits. It was Barrell's view (which is still accepted with minor qualifications) that much of this red sandstone had been laid down, not in quiet bodies of water such as lakes, estuaries, or inland seas, but in wide, torrential rivers draining the great mountain ranges of the Devonian continent. In a climate marked by alternation of extreme dry seasons and torrential rains, the uplands were subject to rapid erosion, and the mud, sand, and gravel produced by their decay were deposited in the river basins at a rapid rate. Periodic exposure, permitting prolonged aeration, oxidized the iron-bearing sediments to produce the typical red color, which is imparted by iron oxide. Such lakes as might appear at flood time in the flatter basins were shallow and inconstant, and during the dry season shrank to stagnant pools or dry mud flats.

With Chamberlain, Barrell believed that the early Devonian ostracoderms, sharks, and ganoid fishes were confined to the continental rivers and lakes; in the Middle Devonian, competition and increasing aridity drove the sharks into the sea and gave over the domination of the continental waters to the ganoids. In answer to the question of why the ganoids could survive where the sharks could not, Barrell proposed that it had been among these fresh-water fishes, and in direct consequence of the periodic contraction of the continental waters, that aerial respiration and its sequel of terrestrial life had been evolved.

The fresh-water theory failed to attain acceptance, however, or even serious consideration, partly because only a tentative analysis of the fossil record had been presented by its adherents. In 1923 such an analysis was attempted by the botanist MacFarlane {36}, who came to the same conclusion as Chamberlain and Barrell; but his argument suffered from the fact that his opinion could not carry the weight of one experienced in either the nature of geological formations or the invertebrate faunal associations, and lacked the authority requisite to the final interpretation of a question capable of being warmly debated between paleontologists. Moreover, he approached the problem with the conviction that the vertebrates had been evolved from fresh-water nemertean worms, and it would appear that in some instances he was led by this a priori bias to interpret the data in a somewhat arbitrary manner.

In 1930, the writer {46} suggested that the glomerular

kidney was an evolutionary adaptation to fresh water and, with E. K. Marshall, Jr. {37} presented extended evidence on this point, derived from the comparative anatomy and physiology of the vertebrate kidney; and, in 1932, in reviewing the problem of water regulation in the teleost and elasmobranch fishes, he advanced new physiological evidence in support of the fresh-water theory {47}.

Paleontologists who remained unconvinced of the correctness of Chamberlain's theory were favorably influenced when, in 1935, A. S. Romer, one of America's outstanding paleontologists, and his colleague, B. H. Grove, presented a detailed re-examination of the habitat problem {45}. Romer and Grove considered *in extenso* only the records of North America up to the close of the Devonian period, but they included brief résumés of the evidence from Europe and elsewhere, and their conclusions were entirely in favor of the fresh-water theory.

This theory has again been challenged, however, by J. D. Robertson {39} of the University of Glasgow, and R. H. Denison {31} of the Chicago Natural History Museum. Robertson relies heavily on the voluminous literature, much of it of an older date, dealing with the Ordovician and Silurian forms, to which he adds the well known argument that the three living protochordate groups, Hemichordata, Urochordata and Cephalochordata, as well as the Myxinoidea, are marine. More cogent, perhaps, are the criticisms of Denison, who has sifted all the available evidence and who concurs in the theory of a marine habitat for the early vertebrates. Romer {43, 44}, though still emphasizing the role of the armor as protection against the eurypterids, continues to adhere to the fresh-water theory.

To these weighty opinions I can add little. My paleontologist friends point out that the problem is a most difficult one, perhaps subject only to a Scottish verdict, 'not proved either way,' and that the physiologist will have to proceed on his own at the present time. As a physiologist, I cannot see how the comparative anatomy of the kidney, the urea-retention habitus of the elasmo-branchs, and related matters can be reconciled with the belief that the vertebrates acquired their definitive ana-tomical and biochemical pattern in salt water, and my preference for the time being therefore remains with the fresh-water hypothesis.

In identifying the Charnian with the Grand Canyon disturbance in North America, we are following Schu-chert and Dunbar {17, p. 90 f}.

The hypothetical structure of the protovertebrate given here follows closely that visualized by Chamber-lain {30}.

Where the invertebrates had armored themselves with shells made of chitin or calcium carbonate, the ostra-coderm armor was composed of calcium phosphate, laid down by fibrous tissue and around vascular channels in concentric layers.

A. S. Romer, who proposed the eurypterid theory of the origin of this armor {40, 42}, accepts fresh water as the habitat of the ostracoderms, and admits that the glomerular structure of the vertebrate kidney is a cogent argument in favor of the fresh-water origin of the verte-brates as a phylum, but he dismisses the fresh-water ori-gin of armor on the grounds that, 'As any section of the plates and scales of the ostracoderms will show, the thick and relatively impervious part of the armor lay beneath a superficial porous region close to the surface in which there was a rich blood supply—a type of construction which would destroy most of the value of the armor under such a theory.' {42, p. 46.} But as Bryant {29} describes the armor of *Astraspis*, it consisted of small, tuberculated plates, each of which contained three zones: an inner, very thick, laminated zone penetrated by verti-cal vascular canals; a middle, dense zone in which the vertical extensions of the underlying vascular canals, now comparatively few in number, were surrounded by

remarkably thick, concentric walls; and an outer layer
of dense material laid down in laminae but penetrated
by radiating fibrous tissue, and capped superficially by
a layer of thick, enamel-like substance comprising the
tip of the tubercule. To these may have been added a
somewhat horny epithelial layer covering the enamel
caps and corresponding to the horny outer layer of the
later fishes and higher vertebrates {33, p. 102}. Bryant's
description does not add up to the destruction of the
value of the armor with respect to osmotic insulation,
but, on the contrary, to its inclusion as the impervious
covering of a soft-bodied animal that was capable of
growth and had to rebuild its armor as it grew.

It may be emphasized that the eurypterids, co-inhabitants of fresh water with the ostracoderms, were
also armored, though with a chitinous shell rather than
bony plates, and it can scarcely be argued that their
armor served to protect them against the bottom-living,
sluggish, jawless vertebrates. However, it may be supposed that, as in the case of the vertebrates, their successful invasion of fresh water was dependent on their
waterproofing. Since mutation must precede adaptation,
it is specious to argue whether the first steps toward the
development of armor occurred while the ancestors of
the vertebrates were still bona fide inhabitants of salt
water, or later when they were seeking squatters' rights
in the brackish lagoons. Escape from the osmotic argument cannot be sought with profit in the fact that even
the marine Cambrian invertebrates (including the trilobites, ancestral to the eurypterids) had sought protection
against predatory enemies or the vicissitudes of life
within calcareous or chitinous shells, because shifting
the origin of armor backward in time and phylogeny
merely emphasizes again that all evolution is preadap-
tive: the favorable mutation always appears in at least
a potential form before, under the pressure of selection,
it proves to be advantageous for survival. The incon-

testable point is that the early vertebrates won the fresh-water habitat because they were waterproofed.

Berrill {28} sees in *Amphioxus* a degenerate form of the pre-Cambrian larval tunicates, which unlike the protovertebrate, returned to the sea and lost its head. He considers it in no way ancestral to the vertebrates 'and not even a satisfactory vertebrate type.'

In arguing against the eurypterid theory the writer has drawn on Schuchert and Dunbar {17, p. 193}, and Gregory {33, p. 62 f}. The eurypterids illustrated in Figure 4 are redrawn from Schuchert and Dunbar.

In 1946 White {48} described an ostracoderm from the late Silurian of Lanarkshire, which he named *Jamoytius kerwoodi*. The two known specimens show no trace of either endoskeleton or exoskeleton, although the segmental muscles are remarkably well preserved. *Jamoytius* had long horizontal fin folds, a long, spineless, dorsal fin, and a short anal fin. White considers that it comes close to the ideal 'chordate' ancestor, which he believes was an unarmored form. Gregory {33, p. 105}, however, suggested that *Jamoytius* may be intermediate between the anaspid ostracoderms and *Amphioxus*, and hence, by implication, secondarily naked. We would supplement Gregory's interpretation by remarking that a single 'naked' ostracoderm carries little weight against the large numbers of armored or heavily scaled forms that are now known and it cannot, in the light of the present evidence, controvert the generally accepted thesis that armor is a basic and primitive character of the earliest vertebrates.

IV. THE KIDNEY

49. BRIDGE, T. W. *Cambridge Natural History: Fishes*. Macmillan Company, London, 1922.

50. CUSHNY, A. R. *The Secretion of the Urine*. Longmans, Green and Company, Ltd., London, first edition, 1917.

51. FRASER, E. A. The development of the vertebrate excretory system. *Biological Reviews*, 25: 159. 1950.

52. MARSHALL, E. K., JR. The secretion of urine. *Physiological Reviews*, 6: 440. 1926.

53. MARSHALL, E. K., JR. The comparative physiology of the kidney in relation to theories of renal secretion. *Physiological Reviews*, 14: 133. 1934.

54. PROSSER, C. L., D. W. BISHOP, F. A. BROWN, JR., T. L. JAHN, AND V. J. WULFF. *Comparative Animal Physiology*. W. B. Saunders Company, Philadelphia, 1950.

55. RAMSAY, J. A. The site of formation of hypotonic urine in the nephridium of *Lumbricus*. *Journal of Experimental Biology*, 26: 65. 1949.

56. RICHARDS, A. N. Physiology of the kidney. *Bulletin of the New York Academy of Medicine*, 14: 5. 1938. Second Series.

57. RICHARDS, A. N. Processes of urine formation. *Proceedings of the Royal Society of London*, B126: 398. 1938.

58. ROBERTSON, J. D. Ionic regulation in some marine invertebrates. *Journal of Experimental Biology*, 26: 182. 1949.

59. SMITH, H. W. The composition of the body fluids of elasmobranchs. *Journal of Biological Chemistry*, 81: 407. 1929.

60. SMITH, H. W. The composition of the body fluids of the goosefish (*Lophius piscatorius*). *Journal of Biological Chemistry*, 82: 71. 1929.

61. SMITH, H. W. The inorganic composition of the body fluids of the Chelonia. *Journal of Biological Chemistry*, 82: 651. 1929.

62. SMITH, H. W. *The Kidney: Structure and Function in Health and Disease*. Oxford University Press, New York, 1951.

'Salt and Water' by M. B. Strauss is quoted with permission from Welt, L. G., *Clinical Disorders of Hydra-*

tion and Acid-Base Equilibrium. Little, Brown and Company, Boston, 2nd ed., 1959.

Unicellular organisms and very small plants and animals that live in rivers and lakes containing only traces of salts do so by virtue of the regulatory powers of the exposed, bounding membranes, aided by subsidiary cellular devices such as contractile vacuoles. These vacuoles are 'contractile' in the sense that they fill and empty periodically as they accumulate water (and perhaps waste products) in the form of a droplet which is then bodily extruded from the cell. Such vacuoles serve to maintain both the volume and the composition of the cell, and they represent the most primitive type of excretory operation. Contractile vacuoles are present in all fresh-water protozoa, and in some species the amount of fluid thus baled out of the body in a few minutes may equal the body weight. They are generally absent in marine forms, but exposure of marine forms to fresh water or diluted sea water may bring vacuoles into operation *de novo* or, if pre-existing, speed up their excretory activity. But the protovertebrate, no matter along what lines it is reconstructed, was too big and complex to rely on this device.

At a higher level, many invertebrates of typically marine nature have invaded brackish or fresh water: a few coelenterates and, in order of increasing numbers of species, mites, clams, crabs, crayfish, annelid worms, and snails; all of them have faced the plethora-of-water problem, and have solved it with varying degrees of success in a variety of ways: by actively absorbing salts (notably sodium chloride) through the respiratory membranes, or by incorporating contractile vacuoles for the excretion of water into a compacted excretory organ. But where osmoregulation has attained a fairly advanced degree of control, as in the fresh-water crayfish, the entire body, except for the oral and respiratory membranes, is covered with waterproof insulation preadaptively evolved in their marine ancestors. As Prosser *et al.* {54} sum up the situation for the invertebrates, 'Osmotic regulation . . . comprises a group of very labile characters. Evolution of [the control of] osmotic function has proceeded in many directions by many small changes.'

The osmotic relations of the invertebrates are summarized by Prosser, *et al.* {54}, and recent data have been added by Robertson {58}.

Reinterpretation of the structure and origin of the vertebrate kidney was foreshadowed by the writer in 1930 {46}, as a result of studies on the regulation of the composition of the blood in the elasmobranch and teleost fishes. These studies indicated that the glomerular, fresh-water fish kidney is the older and primitive type, and that the aglomerular kidney of the marine fishes is a secondary specialization related to marine life. Pursuing the implications of this view, Marshall and the writer {37} presented a detailed re-examination of the structure of the kidney in the fishes, Amphibia, reptiles, birds, and mammals, and concluded that the glomerulus was evolved as an adaptation to the fresh-water habitat of the early vertebrates. The story of the evolution of the structure and function of the vertebrate kidney then began to unfold from the complementary pages of anatomy, physiology, and paleontology—a story that contained many unanticipated implications with respect to vertebrate evolution as a whole.

Nothing like the vertebrate glomerulus is found in the excretory organ of any invertebrate, the nearest approach being the nephridium of the earthworm, *Lumbricus* {55}. To return to the ancestry of the vertebrates and the artificial 'phylum' Chordata, no one has demonstrated any specialized excretory organ in the Hemichordata (balanoglossids) {49, p. 15}, in which excretion apparently is a generalized function of the gut and skin; in the Urochordata (ascidians or tunicates), excretion may be carried out by a mass of clear-walled cells associated with the lower intestine and rectum and derived from the coelom, and it is thought that uric acid and possibly other metabolites accumulate in these cells, never to be discharged during the life of the animal {49, p. 54 f}. True excretory organs are developed only in the Cephalochordata: in *Amphioxus* these take the form of many (about 100) pairs of complex nephridia lying at the sides of the dorsal coelomic canals above the pharynx {49, p. 125 f}. The nephridia of *Amphioxus* are obviously complicated secretory organs and have no similarity to the glomerular-tubular nephron of the vertebrates.

Biologists who see in *Amphioxus* the prototype of the earliest vertebrates pass lightly over the fact that the vertebrate nephron could not by any conceivable stretch of the imagination be derived from Amphioxus's nephridial complex, while those who see in *Amphioxus* either a degenerate, simplified ostracoderm {33, p. 84 f} or the permanent larva of some early vertebrate type {15, p. 20} equally overlook the fact that these nephridia could not by a stretch of the imagination be derived from the vertebrate nephron. Moreover, these excretory organs are of ectodermal rather than mesodermal origin, and all the nephridial elements are confined to the pharynx or head (which by the ardent homologizers is homologized with the gill arches of the vertebrates). Romer {15, p. 27} has said that, 'To make a vertebrate out of an arachnid, the supposed ancestor must have lost almost every characteristic feature he once possessed and reduced himself practically to an amorphous jelly fish before resurrecting himself as a vertebrate.' Considering the definitive nature of the glomerular nephron in all vertebrates (with the obvious exception of the aglomerular marine fishes), these discrepancies in the excretory system (despite all other purported homologies with respect to 'branchial baskets,' 'notochords,' and the like) are enough to reduce the *Amphioxus* theory to a jelly in all discussions of the origin of the vertebrates, and to leave that question as open as it was a century ago.

That the primitive coelomic membrane possessed excretory powers is indicated by the role of this membrane in the invertebrates, by vertebrate embryology, and by the secretory properties of the pericardial and perivisceral membranes in the fishes and reptiles, in which these membranes are derived from the coelom {59, 60, 61}.

For the embryonic structure of the kidney in the fishes see Bridge {49, p. 398 f}. An excellent summary of the function of the archinephric duct is given by Romer {15, p. 389}, and the anatomy of the renal-portal system is described by Fraser {51}.

The reproductive cells continue to be discharged into the coelomic cavity in the most primitive living fishes, the cyclostomes, and escape from the body through a pair of abdominal pores which open into the cloaca. Although other routes of egress for the reproductive cells have been established, abdominal pores persist in most elasmobranch fishes and in the lungfishes, *Polypterus*, the sturgeons, the paddlefish, the gar pike and bowfin —all primitive fishes—but rarely in the Teleostei (bony fishes). These pores may reflect the primitive mode of reproduction, or they may reflect the primitive excretory role of the coelomic cavity {49, p. 402}.

Another primitive feature in the kidney is the persistence in some sharks, the bowfin (*Amia*), and a few Amphibia (the salamanders and a few frogs and toads) of open coelomostomes draining into the renal tubules. More generally, however, such coelomostomes drain into the renal venous sinuses or the renal vein. In both instances they serve, in the manner of lymphatics, to return coelomic fluid to the circulation. Though these persistent coelomostomes may be conceived as a primitive feature, their function is now secondarily diverted to nonexcretory purposes.

For the history of the development of modern renal physiology the reader must be referred to technical sources {50, 52, 53, 56, 57, 62} and to references appended to Chapter X. The comparative anatomy of the kidney is discussed in Chapter VIII.

It has been noted that the energy for glomerular filtration is supplied by the heart and transmitted to the glomeruli through the pressure of the arterial blood. We may therefore think of the glomeruli as playing a passive role in the process of filtration, and no increased burden of work devolves upon them when the concentration of any substance in the plasma is increased to any level whatsoever. But the many processes of tubular reabsorption and excretion are, with few exceptions, 'active' processes: the tubule cells must remove the substance from a low concentration in one medium (either urine or blood) and transport it to the other medium against a concentration gradient, an operation that will not proceed spontaneously but requires the local expenditure of energy by the tubule cells. This energy is made available within the cells by the metabolism of suitable fuel stuffs and fed into the 'transport mechanism' by elaborate enzyme systems, and these circumstances impose upon every reabsorptive or excretory process certain quantitative limitations that take the form of maximal rates of transport. In many cases these maximal rates can be quite accurately measured by presenting to the tubules an excess of any substance by artificially raising its concentration in the plasma, and they prove to be fairly reproducible in successive examinations in any one animal. In general, saturation of one reabsorptive system (for example glucose) does not interfere with the other reabsorptive systems (sodium, potassium, phosphate, sulfate, vitamin C, and so on), showing that each transport system operates more or less independently of all others.

It is not known how many truly independent reabsorptive systems are involved in the vertebrate kidney (or even in man), but the number must be very large. In tubular excretion, however, only two transport systems have been clearly identified; each of these operates on a group of chemically related compounds, one group on nonmetabolizable aromatic acid residues containing a free acetyl group ($-CH_2$ COOH), the other on nonmetabolizable quaternary ammonium bases {62}.

V. THE ELASMOBRANCHS

63a. COHEN, J. J., M. A. KRUPP, AND C. A. CHIDSEY III. Renal conservation of trimethylamine oxide in the spiny dogfish, *Squalus acanthias*. *American Journal of Physiology*, 194: 229. 1958.

63b. COHEN, J. J., M. A. KRUPP, C. A. CHIDSEY III, AND C. I. BILTZ. Effect of trimethylamine and its homologues on renal conservation of trimethylamine oxide in the spiny dogfish, *Squalus acanthias*. *American Journal of Physiology*, 196: 93. 1959.

63c. GRAFFLIN, A. L., AND R. G. GOULD, JR. Renal function in marine teleosts: II. The nitrogenous constituents of the urine of sculpin and flounder, with particular reference to trimethylamine oxide. *Biological Bulletin*, 70: 16. 1936.

64. KEMPTON, R. T. Studies on the elasmobranch kidney: I. The structure of the renal tubule of the spiny dogfish (Squalus acanthias). *Journal of Morphology*, 73: 247. 1943.

65. KROGH, A. *Osmotic Regulation in Aquatic Animals.* Cambridge University Press, 1939.

66. MOY-THOMAS, J. H. *Palaeozoic Fishes.* Chemical Publishing Co., New York, 1939.

67. NEEDHAM, J. Contributions of chemical physiology to the problem of reversibility in evolution. *Biological Reviews*, 13: 225. 1938.

68. SMITH, H. W. The absorption and excretion of water and salts by the elasmobranch fishes. II. Marine elasmobranchs. *American Journal of Physiology*, 98: 296. 1931.
69. SMITH, H. W. The retention and physiological role of the urea in the Elasmobranchii. *Biological Reviews*, 11: 49. 1936.
70. SMITH, H. W., AND C. G. SMITH. The absorption and excretion of water and salts by the elasmobranch fishes. I. Fresh-water elasmobranchs. *American Journal of Physiology*, 98: 279. 1931.
71. WHITE, E. G. A classification and phylogeny of the elasmobranch fishes. *American Museum Novitiates*, No. 837, April, 1936.
72. WHITE, E. G. Interrelationships of the elasmobranchs with a key to the order Galea. *Bulletin of the American Museum of Natural History*, 74: II, p. 25. 1937.
73. YOUNG, L. Z. *The Life of Vertebrates*. Clarendon Press, Oxford, 1950.

In the surviving marine remnants of the ostracoderms, the hagfishes (*Myxine*) and lampreys (*Petromyzon*), the blood is essentially isosmotic with sea water; either the ancestral forms assumed a salt-water habitat before mechanisms were available for controlling the composition of the body fluids in salt water, or else out of 400-odd million years natural selection has left us only two remnants which gave up the fight and took the osmotically easiest way {47, 65, p. 119}. Our classification of the ostracoderms follows that of Gregory {33} and Romer {41, 42}. The evolution of fins from the spines of the ostracoderms is discussed by Gregory and Raven {34}.

The cartilaginous fishes (Chondrichthyes) are divided by some writers into two subclasses, the Elasmobranchii and the Holocephali (Chimaerae) {42}, while others {49, 73} include both in the class Elasmobranchii. The latter classification seems preferable because the modern *Chimaera collei*, *C. monstrosa* and *Callorhynchus millii* share with the sharks, rays, and skates the urea-retention habitus and the dependent mode of reproduction by internal fertilization {69}; these characters are so distinctive as to override the presence or absence of an operculum and other features that have hitherto been the basis of taxonomic separation. If fusion into one class is to be made, the term Elasmobranchii, anatomically descriptive of the Holocephali as well as of the other cartilaginous fishes, is to be preferred because the cartilaginous state (on which the name Chondrichthyes is based) is now accepted to be secondary, whereas the gill structure (*elasmos* = plate; *branchia* = gill) is primitive {49}.

The chief metabolic waste-product requiring excretion in all animals is the nitrogenous end-product formed in the combustion of protein. All the available evidence warrants the conclusion that throughout the history of the vertebrates up to the evolution of the reptiles this end-product has been urea, the most diffusible and non-toxic nitrogenous substance known. However, another nitrogenous metabolite, trimethylamine oxide, is present in substantial amounts in the body fluids and urine of the elasmobranchs and some marine teleosts; in the dogfish it is actively reabsorbed by the renal tubules and contributes significantly to the osmotic pressure of the blood {63a, 63b, 63c, 69}.

The role of urea in the elasmobranch fishes is discussed by Smith {59, 68, 69, 70}.

The studies of J. W. Burger on the secretion of the rectal gland of the dogfish, *Squalus acanthias*, have not been published at the time of writing. The urine in freshly and carefully caught dogfish is distinctly more hypotonic to the plasma than is suggested by the published figures, which are based on animals generally caught on line trawls and hence restrained on the hook for some hours, and subsequently kept in live cars for indefinite periods. In such freshly caught animals, the urine sodium concentration may considerably exceed that of the plasma.

The rectal gland of the elasmobranchs should not be confused with what this writer {59} some years ago called 'Marshall's gland,' which is appendicial to the genitourinary system (rather than the intestine), and which has been described only in some (perhaps not all) Batoidei (skates). The secretion of Marshall's gland has a high concentration of sodium bicarbonate which apparently protects the sperm from the acid urine (the bladder in the female empties into the top of the uterus). The highest recorded bicarbonate concentration for *Raja stabuloforis* is 115 millimols per liter {59}, but Dr. Thomas H. Maren, working at Salisbury Cove in 1959, has found in *R. ocellata* a concentration of 322 millimols per liter (pH 10)—by a considerable margin, the highest value known in any secretion.

Kempton {64} has challenged the older statement that the elasmobranch tubule contains a unique segment. In the spiny dogfish, *Squalus acanthias*, he finds only the typical 'proximal' and 'distal' portions, but it is not certain that these are truly homologous with the corresponding segments in the Amphibia and mammals.

The term 'ovoviviparous' means producing eggs that have a well-developed shell as in oviparous animals, but which hatch within the body of the parent, as in the case of many elasmobranchs and reptiles. The exceptions to ovoviviparity among the elasmobranchs are the orders Chimaeroidei and Cestraciontes (and in the order Euselachii, the families Scylliorhinidae and Hemiscyllidae); in the order Tectospondyli, the family Somniosidae; and, in the order Batoidei, the family Raiidae. The order Cestraciontes includes among its living representatives the rare primitive sharks, *Chlamydoselachus* and *Heterodontus*. All the Euselachii are viviparous except the Scylliorhinidae and Hemiscyllidae, which include the common western dogfishes, *Scyllium canicula* and *S. catulus,* and the eastern *Chiloscyllium griseum* and *C. indicum.* The Greenland shark,' *Laemargus borealis* (family Somnioscidae), is allegedly unique in producing eggs devoid of a horny covering, which are deposited on the sea bottom and fertilized externally (?!) {49, pp. 432 f, 469 f; 69; 72, p. 96}.

The only examples of viviparity among the bony fishes occur in certain families of the Blenniidae (*Zoarces*), the Cyprinodontidae (*Gambusia, Anableps*), the Scorpaenidae (*Sebastes*), the Comephroidae (*Comephorus*), and the Embiotocidae (Surf-fishes). {49, p. 418.}

It is of interest that a few teleosts are hermaphroditic and self-fertilizing (*Serranus cabrilla, S. hepatus* and *S. scriba* and *Chrysoprys auratus*), while hermaphroditism may occur as an occasional variation in the cod, mackerel, and herring {49, p. 420}.

According to MacFarlane {36}, the Permian Ctenacanthii and the Triassic hybodont sharks were restricted to fresh water, and it is not until the Jurassic that the latter begin to appear in marine deposits. Consequently he believed that all elasmobranch evolution proceeded in fresh water until Jurassic time. But it may be that the Triassic forms, and indeed the ctenacanths and hybodonts that carried on through the Permian stricture, were euryhaline, as are certain species within the recent genera *Carcharhinus, Sphyrna, Mustelus, Squalus, Pristis, Dasyatis, Raja,* and others {69}.

In respect to the phylogeny and habitat of the Paleozoic elasmobranchs, we have followed Moy-Thomas {66}, White {71, 72}, Romer and Grove {45}, and Romer {42}.

A list of recent fresh-water elasmobranchs is given by Smith {69}.

That the reduction in blood-urea content of fresh-water elasmobranchs is not a 'reversal of evolution' is stated as a rebuttal to Needham {67}.

VI. THE LUNGFISH

74. ATZ, J. W. Narial breathing in fishes and the evolution of internal nares. *Quarterly Review of Biology,* 27: 366. 1953.
75. BALDWIN, E. *An Introduction to Comparative Biochemistry.* Macmillan Company, New York, 1937.
76. FLORKIN, M. *Biochemical Evolution.* Translated by S. Morgulis, Academic Press, Inc., New York, 1949.
77. ROMER, A. S. The braincase of the carboniferous crossopterygian *Megalichthys nitidus. Bulletin of the Museum of Comparative Zoology* (Harvard), 82: 3. 1937.
78. ROMER, A. S. AND E. C. OLSON. Aestivation in a Permian lungfish, *Breviora. Museum of Comparative Zoology,* 30: 1. 1954.
79. SMITH, H. W. The excretion of ammonia and urea by the gills of fish. *Journal of Biological Chemistry,* 81: 727. 1929.
80. SMITH, H. W. Metabolism of the lung-fish, *Protopterus aethiopicus. Journal of Biological Chemistry,* 88: 97. 1930.
81. SMITH, H. W. Observations on the African lungfish, *Protopterus aethiopicus,* and on evolution from water to land environments. *Ecology,* 12: 164. 1931.
82. SMITH, H. W. *Kamongo.* Viking Press, Inc., New York, 1932. Revised edition, 1949. *Kamongo, or the Lungfish and the Padre.* Compass Books, Viking Press, Inc., New York, 1956.
83. SMITH, H. W. The metabolism of the lung-fish. I. General considerations of the fasting metabolism in active fish. *Journal of Cellular and Comparative Physiology,* 6: 43. 1935.
84. SMITH, H. W. The metabolism of the lung-fish. II.

Effect of feeding meat on metabolic rate. *Journal of Cellular and Comparative Physiology,* 6: 335. 1935.

Romer {77} erected the subclass Choanichthyes (*choana* = funnel; *ichthys* = fish) to include the Dipnoi and Crossopterygii, this name being based upon the presence of internal nostrils (funnels) leading from the mouth to the exterior, a feature that was once supposed to be a sign of aerial respiration. This taxonomic category is equivalent to Hubbs's earlier Amphibioidei, but neither includes the air-breathing, Paleozoic Osteichthyes. There is, therefore, no satisfactory taxonomic category to include all the air-breathing fishes, both Paleozoic and recent. Atz {74} has recently shown that none of the surviving lungfishes uses the internal nostrils for this purpose; rather they gulp air through the mouth, as do many other fishes that supplement their gills with aerial respiration but do not possess internal nostrils. The passages serve only as olfactory organs and there is no reason to suppose that they served a respiratory function in the Devonian fishes. In the recent Amphibia the internal nostrils are used for respiration, but Atz, citing an opinion of Romer's, believes that this respiratory function was acquired in the transitional Crossopterygii or early Amphibia.

Studies on the lungfish, *Protopterus aethiopicus,* have been reported by the writer {80, 81, 83, 84}, and the circumstances attending the collection of the fish in Africa are narrated in *Kamongo* {82}.

In the Paleozoic ganoid fishes, the scales, only a step removed from the armor of the placoderms, were composed of an inner layer of bone and an outer layer of an enamel-like substance, ganoin, and perhaps the close mosaic of these scales precluded the presence of the mucous glands, so important in the estivation of *Protopterus. Lepidosiren,* which estivates, secretes mucus as abundantly as does *Protopterus* but allegedly does not form a cocoon {74}. On the mucous glands of *Neoceratodus*, which does not estivate, the writer has no information. Perhaps the mucus secreted by *Protopterus* is unique in its capacity to form a parchment-like cocoon.

With respect to the statement that *Neoceratodus* dies if removed from water to air, one wonders if it would die if kept in a moist atmosphere; certainly *Protopterus* will die in dry air unless enveloped by the cocoon, and the failure of *Neoceratodus* to estivate does not prove the inadequacy of its lungs as respiratory organs.

The origin of excreted ammonia in the lower animals is obscure. In the course of protein metabolism in the mammal, nitrogen is first degraded to ammonia, but this compound is relatively toxic and it is converted by the liver into the neutral and nontoxic urea. Such ammonia as is excreted in the urine of mammals in the maintenance of acid-base balance is not that produced directly by protein metabolism, but ammonia that is manufactured *de novo* from glutamine and other amino acids by the renal tubules and added to the urine on its way out of the body {62, p. 401}.

The recorded data {54} show that many invertebrates excrete their nitrogen chiefly in the form of ammonia, and they have consequently been called 'ammonotelic' on the assumption that this predominance of ammonia is indicative of a unique type of protein metabolism, distinguishing them from 'ureotelic' (urea-excreting) forms. However, the possibility that this ammonia may be formed peripherally and facultatively has been wholly ignored by students of invertebrate physiology, and we believe that the characterization 'ammonotelic' is unwarranted in many instances, and sometimes misleading— as for example in the earthworm, which excretes ammonia when fasting and urea when fed.

In the fishes considerable ammonia is excreted by the gills as well as the kidneys. In our initial study of nitrogen excretion by marine fish {79}, we concluded that the branchially excreted ammonia was derived directly from blood ammonia, presumably by diffusion; but subsequent studies in the lungfish {80} led us to abandon this view in favor of peripheral formation, and in the study of the fresh-water elasmobranch, *Pristis microdon* {69, 70}, evidence was obtained that the branchial excretion of ammonia is under physiological control.

The statement that ammonia excretion is characteristic of all aquatic organisms {75}, and the implication that this ammonia is the ultimate, hepatic product of protein metabolism in the fresh-water and marine teleosts and the lungfish {76}, ignore both the abundant branchial excretion of ammonia and urea, as compared with urinary excretion, and the possibility of peripheral ammonia formation.

Leon Goldstein and R. P. Forster, working at Salisbury Cove, have recently (1959) shown that the ammonia excreted by the gills of the marine shorthorn sculpin, *Myoxocephalus scorpius,* is formed *de novo.* (Only small amounts of urea are excreted by the gills in this species, and urea excretion in the urine is probably negligible.) The gill is rich in glutaminase and glutamic dehydrogenase, the total maximal activity of which is about equivalent to the observed ammonia excretion. The presence of these enzymes indicates that glutamine is a major precursor of the branchial ammonia in this fish, as it is in the mammalian kidney.

In the frog and mud puppy ammonia is practically absent from the blood and glomerular filtrate, but is secreted into the urine by the distal tubule {62}, and it may confidently be accepted that nearly all urinary ammonia in the Amphibia is of peripheral origin.

The invertebrates aside, until cogent evidence to the contrary is presented, we propose, on the evidence supplied by the elasmobranch and teleost fishes, the lungfish, the frog, and the mud puppy, that the primary nitrogenous end-product of protein metabolism in the vertebrates, with the exception of the uric-acid-excreting reptiles and birds, is urea, and that ammonia formation is peripheral and probably facultative; and we conceive that such was the case in the Crossopterygii and Paleozoic Amphibia.

VII. THE AMPHIBIA

85. DAWSON, A. B. Functional and degenerate or rudimentary glomeruli in the kidney of two species of Australian frog, Cyclorana (Chiroleptes) playtcephalus and alboguttatus (Günther). *Anatomical Record,* 109: 417. 1951.

86. FORSTER, R. P. The use of inulin and creatinine as glomerular filtrate measuring substances in the frog. *Journal of Cellular and Comparative Physiology,* 12: 213. 1938.

87. FORSTER, R. P. The nature of the glucose reabsorptive process in the frog renal tubule. Evidence for intermittency of glomerular function in the intact animal. *Journal of Cellular and Comparative Physiology,* 20: 55. 1942.

88. GLADSTONE, R. J., AND E. P. G. WAKELEY. *The Pineal Organ.* Williams & Wilkins Co., Baltimore, 1940.

89. GRAFFLIN, A. L., AND E. H. BAGLEY. Glomerular activity in the frog's kidney. *Bulletin of the Johns Hopkins Hospital,* 91: 306. 1952.

90. JØRGENSEN, C. B. Permeability of the amphibian skin. II. Effect of moulting of the skin of anurans on the permeability to water and electrolytes. *Acta Physiologica Scandinavica,* 18: 171. 1949.

91. JØRGENSEN, C. B. The amphibian water economy, with special regard to the effect of neurohypophysial extracts. *Acta Physiologica Scandinavica,* 22: Suppl. 78. 1950.

92. MILLOT, J. New facts about the coelacanths. *Nature,* 174: 426. 1954.

93. MILLOT, J. First observations on a living coelacanth. *Nature*, 175: 362. 1955.

94. MILLOT, J. The coelacanth. *Scientific American*, 193: (6) 34. 1955.

95. PEARSE, A. S. Concerning the development of frog tadpoles in sea water. *The Philippine Journal of Science*, 6: 219. 1911.

96. SAWYER, W. H. Increased renal reabsorption of osmotically free water by the toad (*Bufo marinus*) in response to neurohypophysial hormones. *American Journal of Physiology*, 189: 564. 1957.

97. SCHAEFFER, B. *Latimeria* and the history of coelacanth fishes. *Transactions New York Academy of Sciences*, Series II, 15: 170. 1953.

98a. SMITH, J. L. B. *The Story of the Coelacanth.* Longmans, Green and Company, Ltd., London and New York, 1956.

98b. USSING, H. H. General Principles and Theories of Membrane Transport. In *Metabolic Aspects of Transport Across Cell Membranes*. Q. R. Murphy, Ed. University of Wisconsin Press, Madison, Wis. 1957. See also Proceedings of the XXI International Congress of Physiology (1959).

99. WATSON, D. M. S. The reproduction of the coelacanth fish, *Undina. Proceedings of the Zoological Society of London*, 1: 453. 1927.

In this chapter, as elsewhere, we have followed Schuchert and Dunbar {17} in matters relating to historical geology and paleoclimatology, with slight modifications, with respect to the latter, based largely on Brooks {3}.

The evolution of the tetrapod foot is discussed by Gregory and Raven {34}, Gregory {33}, and Romer {15, 42}.

The discovery of the coelacanth is related by J. L. B. Smith {98} and anatomical observations have been reported by Millot {92, 93, 94}. Schaeffer's article {97} gives a good synopsis of ancient forms.

The pineal eye in extinct and recent vertebrates is discussed by Young {73, pp. 102, 162}, Gregory {33, pp. 338, 526} and Bridge {49}. An older but definitive work on the subject is Gladstone and Wakeley's monograph {88}.

In the higher animals the pituitary gland is a complex structure separable into two major divisions. The preferred name for the gland as a whole is '*hypophysis*' (*hypo* = under; *physis* = nature), based on the fact that it is located 'under' the brain. One division, the *adenohypophysis* (*aden* = gland; + *hypophysis*) contains tissues known as the *pars distalis* (= anterior pituitary), the *pars tuberalis*, and the *pars intermedia*, all of which have an obvious glandular structure. The second division, or *neurohypophysis* (*neuron* = nerve; + *hypophysis*) or neural lobe, is so named because it is composed in great part of nerve fibers, with minor obviously glandular elements {62}.

The adenohypophysis secretes several hormones which promote metabolic processes in the tissues generally and in other glands, and which are therefore 'tropic' (*trope* = adjectival form meaning 'to turn') hormones, of which the growth hormone, the adrenocorticotropic hormone (ACTH), and the thyrotropic hormone are generally familiar. That the adenohypophysis was functionally active in the ostracoderms is indicated by the fact that extracts of the mammalian gland induce changes in secondary sex characters in the cyclostomes, as well as ovulation in the fishes and Amphibia.

The neurohypophysis secretes the antidiuretic hormone (ADH), the oxytocic hormone (which causes contraction of the pregnant uterus), and the amphibian 'water-balance' hormone, while the pars intermedia secretes the melanophore-expanding hormone.

The effective agent reducing the glomerular circulation in the frog appears to be the oxytocic fraction, but in the toad, alligator, and chicken, the effective agent is ADH. With respect to reduction in glomerular activity, the toad, which is better adapted to terrestrial conditions than is the frog, is much more sensitive to neurohypophysial extracts than is the latter animal, indicating that pituitary control of the glomeruli is subject to quantitative variation even among the Amphibia {96}.

In the mammals the oxytocic fraction probably participates in parturition by stimulation of the uterus, and in lactation; what function it may have had in the ostracoderms and fishes lies beyond conjecture.

Whether a pituitary hormone is involved in the control of the glomerular circulation in the fishes, and notably in marine forms where glomerular activity is markedly reduced (Chapter VIII), is undetermined.

Intermittent glomerular activity, involving some or even all the glomerular capillaries, was described many years ago in the frog's kidney by A. N. Richards and C. F. Schmidt, but A. L. Grafflin and E. H. Bagley {89} have recently adduced evidence that this phenomenon is attributable to the abnormal state of the experimental animals, and that spontaneous intermittence is probably not a characteristic feature of the glomerular circulation under normal conditions. This conclusion, however, does not detract from the significance of reduction in glomerular filtration (by alternation of activity or otherwise) during dehydration {86, 87}. The problem is, When is a frog in a normal state of hydration—when submerged to the nose in water, or when sitting on a shaded lily pad?

The absorption of sodium chloride by the skin in the Amphibia has been extensively studied by Krogh {65} whose work is summarized by Prosser et al. {54}, by Jørgensen {90}, and most extensively by Ussing and his collaborators {98a}.

Glomerular development in *Cyclorana* is discussed by Dawson {85}. Water economy in this and related forms is discussed by Buxton (see Chapter XI).

This is as good a place as any to discuss the function of a urinary bladder. Whatever usefulness may be read into this organ in the higher animals, it is rather surprising that a urinary bladder is fully developed in all the Amphibia and in most fishes. No bladder is present in the elasmobranchs, in which the ureters open into a common urogenital sinus or directly into the cloaca. Neither is a bladder present in the cyclostomes or the Dipnoi; but one is present in the surviving ganoids and generally in the teleosts, in which the archinephric ducts unite and expand into a single urinary sinus {49, p. 407}. The bladder in the fishes is, however, derived from mesoderm and is not homologous with the cloacal bladder of the tetrapods. No bladder is present in the reptiles, except for the lizards and turtles, or in the birds or the two primitive egg-laying mammals, the platypus and spiny ant-eater.

In the Amphibia one can see how a large urinary bladder might afford a small supply of dilute fluid to be reabsorbed when the animal is threatened with desiccation, but that this is the explanation of its evolution must be set aside in view of the presence of the organ in the primitive fresh-water fishes, where there is no need for such reabsorption, as well as in many highly specialized marine fishes where water reabsorption would be impossible because the urine is already maximally concentrated. Most marine fishes have a very large bladder and it is frequently found to be full of urine. What seems a more cogent suggestion is that the urine, being perhaps detectable by enemies, could leave an unbroken trail for pursuit were it not retained and discharged at intervals. The goosefish, *Lophius*, for example, may carry around with it as much as 5 per cent of its body weight as urine, though at the normal rate of urine formation several days would be required for this quantity of urine to accumulate. This theory could be extended back to all fishes that are not vigorous and continuous swimmers, and thus would exclude the pelagic sharks (but not the skates and rays), though it would not explain the absence of a bladder in the lungfishes. Consequently we can only say that the evolution of a large urinary bladder in the fishes remains a mystery.

The record of salt-water frogs is that of Pearse {95}.

VIII. THE BONY FISHES

100. ALLEE, W. C., AND P. FRANK. Ingestion of colloidal material and water by goldfish. *Physiological Zoology*, 21: 381. 1948.

101. BERGLUND, F., AND R. P. FORSTER. Renal tubular transport of inorganic divalent ions by the aglomerular marine teleost, *Lophius americanus*. *Journal of General Physiology*, 41: 429. 1958.

102. BIETER, R. N. Further studies concerning the action of diuretics upon the aglomerular kidney. *Journal of Pharmacology and Experimental Therapeutics*, 49: 250. 1933.

103. BIETER, R. N. The action of diuretics injected into one kidney of the aglomerular toadfish. *Journal of Pharmacology and Experimental Therapeutics*, 53: 347. 1935.

104. BREDER, C. M., JR. Ecology of an oceanic freshwater lake, Andros Island, Bahamas, with special reference to its fishes. *Zoologica*, 18: 57. 1934.

105. BURNS, J., AND D. E. COPELAND. Chloride excreting cells in the head region of Fundulus heteroclitus. *Biological Bulletin*, 99: 381. 1950.

106. CLARKE, R. W. The xylose clearance of Myoxocephalus octodecimspinosus under normal and diuretic conditions. *Journal of Cellular and Comparative Physiology*, 5: 73. 1934.

107. CLARKE, R. W. Simultaneous xylose and inulin clearances in the sculpin. *Bulletin of the Mount Desert Island Biological Laboratory*, 25. 1936.

108. COPELAND, D. E. Adaptive behavior of the chloride cell in the gill of Fundulus heteroclitus. *Journal of Morphology*, 87: 369. 1950.

109. DAKIN, W. J. Presidential Address: The aquatic animal and its environment. *Proceedings of the Linnean Society of New South Wales*, 60: Parts 1–2, vii. 1935.

110. EDWARDS, J. G. Studies on aglomerular and glomerular kidneys. I. Anatomical. *American Journal of Anatomy*, 42: 75. 1928.

111. EDWARDS, J. G. Studies on aglomerular and glomerular kidneys. III. Cytological. *Anatomical Record*, 44: 15. 1929.

112. EDWARDS, J. G. The renal unit in the kidney of vertebrates. *American Journal of Anatomy*, 53: 55. 1933.

113. EDWARDS, J. G. The epithelium of the renal tubule in bony fish. *Anatomical Record*, 63: 263. 1935.

114. EDWARDS, J. G., AND L. CONDORELLI. Studies on aglomerular and glomerular kidneys. II. Physiological. *American Journal of Physiology*, 86: 383. 1928.

115a. FORSTER, R. P. A comparative study of renal function in marine teleosts. *Journal of Cellular and Comparative Physiology*, 42: 487. 1953.

115b. FORSTER, R. P., AND F. BERGLUND. Osmotic diuresis and its effect on total electrolyte distribution in plasma and urine of the aglomerular teleost. *Lophius americanus. Journal of General Physiology*, 39: 349. 1956.

115c. FORSTER, R. P., F. BERGLUND, AND B. R. RENNICK. Tubular excretion of creatine, trimethylamine oxide, and other organic bases by the aglomerular kidney of *Lophius americanus. Journal of General Physiology*, 42: 319. 1958.

116. GETMAN, H. C. Adaptive changes in the chloride cells of Anguilla rostrata. *Biological Bulletin,* 99: 439. 1950.

117. GRAFFLIN, A. L. Renal function in marine teleosts. I. Urine flow and urinary chloride. *Biological Bulletin,* 69: 391. 1935.

118. GRAFFLIN, A. L. Renal function in marine teleosts. III. The excretion of urea. *Biological Bulletin,* 70: 228. 1936.

119. GRAFFLIN, A. L. Renal function in marine teleosts. IV. The excretion of inorganic phosphate in the sculpin. *Biological Bulletin,* 71: 360. 1936.

120. GRAFFLIN, A. L. Observations upon the aglomerular nature of certain teleostean kidneys. *Journal of Morphology,* 61: 165. 1937.

121. GRAFFLIN, A. L. Observations upon the structure of the nephron in the common eel. *American Journal of Anatomy,* 61: 21. 1937.

122. GRAFFLIN, A. L. The problem of adaptation to fresh and salt water in the teleosts, viewed from the standpoint of the structure of the renal tubules. *Journal of Cellular and Comparative Physiology,* 9: 469. 1937.

123. GRAFFLIN, A. L. The structure of the nephron in the sculpin, Myoxocephalus octodecimspinosus. *Anatomical Record,* 68: 145. 1937.

124. GRAFFLIN, A. L., AND D. ENNIS. The effect of blockage of the gastro-intestinal tract upon urine formation in a marine teleost, Myoxocephalus. *Journal of Cellular and Comparative Physiology,* 4: 283. 1934.

125. GUYTON, J. S. The structure of the nephron in the South American lungfish, Lepidosiren paradoxa. *Anatomical Record,* 63: 213. 1935.

126. HUBER, G. C. On the morphology of the renal tubules of vertebrates. *Anatomical Record,* 13: 305. 1917.

127. KEOSIAN, J. Secretion in tissue cultures. III. Tonicity of fluid in chick mesonephric cysts. *Journal of Cellular and Comparative Physiology*, **12**: 23. 1938.

128. MACALLUM, A. B. The paleochemistry of the body fluids and tissues. *Physiological Review*, **6**: 316. 1926.

129. NASH, J. The number and size of glomeruli in the kidneys of fishes with observations on the morphology of the renal tubules of fishes. *American Journal of Anatomy*, **47**: 425. 1931.

130. PITTS, R. F. Urinary composition in marine fish. *Journal of Cellular and Comparative Physiology*, **4**: 389. 1934.

131. SMITH, H. W. The excretion of phosphate in the dogfish, Squalus acanthias. *Journal of Cellular and Comparative Physiology*, **14**: 95. 1939.

Some years ago the Canadian physiologist, A. B. Macallum {128}, advanced the view that the blood plasma of the vertebrates and invertebrates with a closed circulatory system is, in its inorganic salts, a reproduction of the sea water of the geological period in which the representatives of such forms first made their appearance. This theory is clearly inapplicable to the vertebrates, which were evolved not in the sea but in fresh water, and it is questionable if it is applicable to any invertebrates. Macallum's thesis has been cogently criticized by W. J. Dakin {109} in an excellent paper that has been generally overlooked by students of this subject.

Of the extensive literature on the anatomy and physiology of the kidney in the lower vertebrates, only the most recent will be included here in order to bring this literature up to date, but those cited will supply a guide to earlier papers. Anatomical studies have been reported by numerous investigators {37, 53, 85, 110, 111, 112,

113, 114, 120, 121, 122, 123, 125, 126, 129}. Physiological studies are reported in 46, 47, 53, 56, 57, 87, 89, 101, 102, 103, 106, 107, 114, 115a–c, 117, 118, 119, 124, 130, 131. Additional literature is cited by Smith {62, pp. 35, 110, 112, 143, 179, 183, 520, 522} and in literature appended to Chapter X.

No observations are available on the filtration rate in the eel, salmon, or stickleback, in relation to migration between fresh and salt water, and relatively few observations are available on fresh water {53} or marine {106, 107, 117} teleosts generally. The longhorn sculpin *M. octodecimspinosus* has been studied extensively by R. P. Forster {115}. The large magnitude of the urine flow in the fresh-water eel, carp, and goldfish {46, 53} implies a correspondingly large filtration rate.

It is clear from the studies of Edwards {112, 113} that the proximal tubule in the mammal, bird, reptile, and frog is cytologically uniform throughout its length. However, in the glomerular nephron of the fresh-water and marine teleosts, Edwards {111, 113} has shown that the proximal tubule is differentiated into two portions (or segments). According to Grafflin {122} the first portion, where present, is homologous in all fish kidneys; but this first portion is lost in the aglomerular kidney, only the second remaining.

Edwards {112, 113} has also shown that in typical marine teleosts, whether the nephron is glomerular or aglomerular, the distal tubule is regularly absent, whereas in the fresh-water teleosts the distal tubule is regularly present. In this respect, the common eel (*Anguilla rostrata*) is a typical fresh-water form, in that it possesses a distal tubule {121}, whereas the stickleback, *Fundulus heteroclitus*, is a typical marine form, in that it lacks a distal tubule {112, 122}. This difference conforms with other evidence that the eel, though it invades the sea, has the congenital features characteristic of a fresh-water habitus, and that the stickleback, though it invades fresh water, has the congenital features characteristic of a marine habitus.

The peritubular blood supply of the aglomerular kidney could conceivably have been derived from the renal artery, but this would have required the evolution of some device (analogous to the glomerulus) to reduce the pressure of this arterial blood before it entered these capillaries; otherwise water would be filtered out of the capillaries into the interstitial space between the tubules, with no way of removing it.

Keys has shown by means of a heart-gill preparation that branchial excretion in the eel can move chloride against a concentration gradient of nearly 3 to 1. Our assumption that ion transport is effected by the undifferentiated respiratory epithelium has been challenged by him and by Copeland and others, who attribute the process to isolated cells that occur not only in the gills but also in other regions of the head, especially on the inner surface of the operculum. Opposed to this interpretation is the belief of Bevelander that these specialized cells are mucous cells and have no special relation to ion transfer. The literature of the subject is cited by Copeland {108}, Burns and Copeland {105}, and Getman {116}. The problem is under further study by several investigators at the present time.

It is theoretically feasible for the marine fish to absorb water (without salt) directly from sea water through the gill membranes, since the same quantity of work would be involved in the separation of salt from water. There are two possible explanations why the circuitous method of drinking sea water is used. It has been noted that in the invertebrates the respiratory epithelium is apparently concerned in the regulation of the ionic composition of the body fluids, even where the animal's blood is isosmotic with sea water {54, p. 92}; hence the excretion of sodium chloride by the gills in both the

elasmobranchs and teleosts is possibly an archaic operation which was inherited by the vertebrates from their
invertebrate ancestors. As an ion transport mechanism,
the gill may have primitively been incapable of osmotic
work with reference to water. Had it been otherwise, one
may ask why the marine elasmobranchs had to turn to
the urea-retention habitus in order to separate water
from sea water, or why the ostracoderms failed to establish themselves in the sea.

An equally cogent consideration, however, is the fact
that to absorb water directly from sea water, the membrane (or cell) involved must operate *specifically* on
water molecules in separating water from salt. Though
many unicellular organisms can pump water *out* of the
cell, there is as yet little evidence that any unicellular
organism, or any cell or organ in any animal, can pump
water *into* the cell against an outward osmotic gradient,
and it may be that this is a trick that protoplasm has
never solved. The mechanism of making a concentrated
urine in the birds and mammals is now interpreted in
terms of the active transport of sodium chloride. (Chapter X.)

The process of drinking sea water, as practiced by the
marine teleosts, is not a very efficient one. Sea water contains, in addition to sodium chloride, considerable quantities of the divalent ions, calcium, magnesium, and sulfate, all of which are poorly absorbed by the intestinal
mucosa (as is demonstrated by the familiar use of magnesium sulfate or Epsom salts as a 'saline' or osmotic
cathartic). As sodium chloride is absorbed, these divalent
ions are in large part left behind in the intestinal residue
and draw water from the body fluids to produce a solution which approaches the blood in osmotic pressure;
on opening the intestine of a marine fish, one finds in it
large quantities of this unabsorbable Epsom salts mixture, and this clear fluid is frequently discharged from
the anus with vigor when the fish is handled. Some calcium, magnesium, and sulfate are, however, absorbed

into the blood and must be excreted in the urine. These divalent ions, with phosphate and some additional sulfate derived from protein metabolism, comprise the major inorganic constituents in the urine, which normally is almost sodium chloride free {124}. So far as water balance is concerned, these divalent ions involve so much waste motion because they require for their excretion more water than accompanies them into the body in the ingested sea water.

This unique urinary mixture in the marine teleosts poses another physiological problem for the animal, because calcium and magnesium salts, when brought to neutrality, precipitate as the corresponding insoluble hydroxides or oxides; and in neutral solution, in the presence of phosphate ions, magnesium forms the insoluble magnesium acid phosphate {130}. Consequently, were the urine to shift from an acid to neutral or alkaline reaction, precipitation of these salts would lead to obstruction of the renal tubules. This consideration is undoubtedly related to the fact that the urine cannot be rendered alkaline in the dogfish {131} or marine teleost by the administration of sodium bicarbonate and other methods which are effective in other animals.

Water is, of course, also excreted by the aglomerular kidney, but it may be assumed that this water is drawn passively into the tubular urine by osmosis as the urinary solutes are excreted. The observations of Keosian {127} on the osmotic pressure of the fluid in *in vitro* cultures of the chick mesonephros support the latter interpretation, as do those of Bieter {101, 102, 103} on the diuresis produced by various salts in the toadfish. We believe the fact that the secretion pressure during salt diuresis in the aglomerular kidney can exceed the pressure in the dorsal aorta may be interpreted as an osmotic phenomenon, rather than as evidence for the specific secretion of water. In this interpretation, the fact that during immersion in diluted sea water the inulin U/P ratio in the

glomerular longhorn sculpin, *M. octodecimspinosus*, may
be less than 1.0 {115}, may similarly reflect increased
ingestion of sea water and increased excretion of magnesium and sulfate.

Our interpretation of the theater of evolution of the
teleosts differs from that of Romer, who believes that
the ancestors of the marine teleosts invaded the sea in
the Carboniferous and that the present fresh-water fishes
generally have reinvaded that medium in recent times
{15, p. 51; 42}.

The literature on the osmotic relations of the teleosts
with respect to their environment is summarized by
Dakin {109, p. vii} and Prosser, *et al.* {54}.

The migration of marine fishes into hard fresh-water
pools is discussed by Breder {104}.

That goldfish ingest water in fresh water has been
demonstrated by Allee and Frank {100}, but the goldfish feeds on microscopic plants and animals, whereas
the eel and many other fresh-water fishes do not.

IX. THE REPTILES AND BIRDS

132. COULSON, R. A., T. HERNANDEZ, AND F. G. BRAZDA.
    Biochemical studies on the alligator. *Proceedings
    of the Society for Experimental Biology and Medi-
    cine*, 73: 203. 1950.
133. KHALIL, F. Excretion in reptiles. *Journal of Bio-
    logical Chemistry*, 171: 611. 1947.
134. MOYLE, V. Nitrogenous excretion in chelonian rep-
    tiles. *Biochemical Journal*, 44: 581. 1949.
135. ROMER, A. S. Origin of the amniote egg. *Scientific
    Monthly*, 85: 57. 1957.

The literature on nitrogen excretion in the reptiles and
birds is summarized by Prosser *et al* {54} and Smith
{62, p. 131, p. 524}, and new data have recently been
added by Moyle {134} and Khalil {133}.

That the reptiles are all characterized by the uric acid habitus has a qualified exception in the turtles or Chelonia, because in aquatic turtles uric acid rarely comprises more than 30 per cent of the total urinary nitrogen, and frequently the figure is less than 10 per cent. Information on the Chelonia is limited and it is appropriate to refer here to some unpublished observations made by the writer in 1925–1928 on fresh-water, terrestrial, and marine turtles. These observations conform with the recorded data in showing that the urinary nitrogen in G. *pseudographica*, E. *blandingii*, K. *subrubrum hippocrepis*, and *Kinosternon sp.*, C. *marginata belli*, P. *conccinna*, P. *elegans*, and C. *serpentina*, is variably distributed between urea, uric acid, and ammonia, with small quantities of creatine and creatinine. However, these studies, like others reported in the literature, were not carried out with proper consideration of the nutritional condition of the animal, nor was consideration given to the possible loss of urea or ammonia through anal sacs.

In the marine turtles, *Caretta kempi*, C. *caretta*, *Chelonia mydas*, and *Eretmochelys imbricata*, examined at the New York Aquarium through the courtesy of Dr. C. M. Breder, only a small fraction (1.7 to 5 per cent) of the urinary nitrogen *in solution* is uric acid, but the urine frequently shows heavy mucilaginous strands that are rich in this substance. Urea accounts for 26 to 85 per cent of the urinary nitrogen in a dozen samples, ammonia for 1.5 to 25 per cent, with creatinine and creatine representing less than 1 per cent each. The urea concentration was highest in animals that were known to have eaten recently. Thus the marine turtles do not appear to differ from fresh-water forms in respect to the distribution of nitrogen in the bladder urine.

A notable feature in these data, and one hitherto overlooked, is that the urea content of the plasma and perivisceral fluid in presumptively healthy specimens of *Caretta kempi* and C. *caretta* ranged from 340 to 400 mg. per 100 cc. (the two fluids being identical in this respect) in fed animals, but had a value of only 40 mg. per 100 cc. in a fasting specimen of *C. kempi*. This high plasma urea concentration may reflect the naturally oliguric state, but excessive tubular reabsorption of urea in marine and other reptiles certainly invites further investigation.

That the marine turtles ingest a little sea water in captivity is indicated by significant concentrations of magnesium and sulfate in the stomach, the concentrations of these ions increasing through the anterior and posterior intestine as the concentration of chloride decreases to negligible values; the urine contains from 10.4 to 69.5 mM. per liter of magnesium, 49.5 to 101 mM. of sulfate, 46 to 56 mM. of chloride, and 30 to 83 mM. of phosphate. However, the intestinal fluid invariably approaches the isosmotic state as it passes down the gastrointestinal tract. The urine is slightly hypotonic to the blood: $\Delta U/\Delta P$ (°C.) $= 0.70/0.76$ in *C. caretta* and $0.60/0.66$ in *C. kempi*. (Δ°C. is the osmotic pressure as measured by freezing point lowering.) ΔU in other specimens of *C. kempi* was 0.67, 0.65, and 0.64, ΔP, 0.77, 0.78, and 0.83. The data do not suggest that *Caretta* drinks sea water as a source of free water, but rather that it handles accidental gulps exactly as would be expected in view of the presence of the nasal salt gland (Chapter XI). We infer that the marine turtles, like the marine birds and mammals, subsist primarily upon metabolic water derived from the food and do not habitually drink sea water.

The order Chelonia ($=$ Testudinata) branched off from the terrestrial reptiles very early, and by the Upper Jurassic had produced marine turtles which may have been descended from fresh-water rather than terrestrial forms, so that the order has possibly had a long fresh-water experience permitting it to become negligent of water conservation. Moreover, many fresh-water diving turtles possess anal sacs into which water is drawn for respiratory purposes when the animal is submerged, and it would not be surprising if urea is excreted by simple diffusion into the fluid contained in these sacs, a circumstance that would favor the urea habitus and that may also have distorted the recorded data on urinary nitrogen. Anal sacs are apparently not present in the marine turtles, but urea may be lost through the oral membranes.

The alligator (A. *mississippiensis*) is likewise a freshwater animal, and likewise an exception to the uricotelic rule; but urea is present, if at all, in the blood and urine only in traces. The chief nitrogenous constituent of the urine is ammonia, which is excreted with large quantities of bicarbonate. But the alligator remains a reptile, and of what nitrogen it does not excrete as ammonia, the greater part is excreted as uric acid {54, p. 194; 132}.

Needham {67} continues in his earlier belief that the form in which nitrogen is excreted depends primarily on the conditions under which the embryo has to live, applying this principle to the elasmobranchs, which, having first developed severe uremia (the reason for the uremia not being stated), were then able to enclose the embryo completely, in order to promote (in an unspecified way) its advanced development. He argues in a parallel manner that, in the birds, uric acid metabolism is an adaptation to the cleidoic (closed) egg, since the accumulation of urea (but not uric acid) would presumably have pathological effects. This interpretation with respect to both the elasmobranchs and the birds (and reptiles) appears to the writer as an inverted reading of the facts, and our earlier reply {69} seems to require no amendment.

To Needham's recent statement that 'No other generalization will explain the lack of uricotelism (uric acid excretion) among mammals, a plainly terrestrial class' {67, p. 237 footnote 1}, it can be replied that no generalization is needed other than biochemical continuity: the fishes, Amphibia, and presumably the nascent amniotes were ureotelic (that is, characterized by urea excretion) and the mammals have simply persisted in this habitus. This issue has been confused as a consequence of the misinterpretation of nitrogen excretion in the fishes, as noted in the last technical note accompanying Chapter VI.

X. THE MAMMALS

136. GRAFFLIN, A. L. The normal, the acromegalic and the hyperplastic nephritic human nephron. *Archives of Pathology*, **27**: 691. 1939.

137. SIMPSON, G. G. The beginning of the age of mammals. *Biological Reviews*, **12**: 1. 1937.

138. SMITH, H. W. *Principles of Renal Physiology*. Oxford University Press, New York, 1956.

139. SMITH, H. W. The fate of sodium and water in the renal tubules. *Bulletin of the New York Academy of Medicine*, 2nd Series, **35**: 293. 1959.

140. SMITH, H. W. Highlights in the history of renal physiology. *The Georgetown Medical Bulletin*, **13**: 4. 1959.

141. SMITH, H. W. The Kidney. In: *The Fabric of Cardiovascular Concepts*. Edited by A. P. Fishman and D. W. Richards. Oxford University Press, New York.

142. SPERBER, IVAR. Studies on the mammalian kidney. *Zoologiska Bidrag Fran Uppsala*, **22**: 252. 1944.

The evolution of the mammals is discussed by Simpson {137}, Romer {41, 44}, Gregory {33}, and Young {73}.

The only reconstructions of entire human nephrons, including the thin segment, are those of L. A. Turley, which have been described by Grafflin {136}.

The comparative anatomy of the kidney as a whole and of individual nephrons in a wide range of mammals is available in Sperber's excellent monograph {142}.

Data on renal function in a variety of mammals, including man, are summarized by Smith {62, p. 529, 138}, and certain aspects of the history of the anatomy and physiology of the kidney have recently been reviewed elsewhere {140, 141}.

The role of the loop of Henle in concentrating the urine has been treated very briefly in this book, but a detailed discussion and history of the subject is readily available {139}. The 'pore' theory of the action of ADH in the mammalian nephron was advanced independently by Wirz and Sawyer {96}.

The power of osmotic concentration is slightly developed in the birds, but not nearly to the extent to which it is developed in the mammals {62, p. 525}. The slight qualitative similarity probably reflects the coincidence of homeothermy and the selective pressure of aridity in both groups. The birds had less reason to evolve the mammalian type of kidney: their problem of water conservation had been solved for them by their reptilian ancestors in the excretion of uric acid; the mammals, excreting their nitrogen as soluble urea, had to develop a concentrating kidney or remain at a great disadvantage in water economy.

XI. ANIMALS THAT LIVE WITHOUT WATER

143. ADOLPH, E. F. (AND ASSOCIATES). *Physiology of Man in the Desert.* Interscience Publishers, Inc., New York, 1947.

144. BABCOCK, S. M. Metabolic water: Its production and role in vital phenomena. *Research Bulletin No. 22, 29th Annual Report, Agricultural Experiment Station, University of Wisconsin,* 1912.

145. BRADLEY, S. W., AND R. J. BING. Renal function in the harbor seal (Phoca vitulina L. during asphyxial ischemia and pyrogenic hyperemia). *Journal of Cellular and Comparative Physiology,* 19: 229. 1942.

146. BUXTON, P. A. *Animal Life in Deserts—A Study of the Fauna in Relation to the Environment.* Edward Arnold & Company, London, 1923.

147. CLARKE, G. L., AND D. W. BISHOP. The nutritional value of marine zooplankton with a consideration of its use as an emergency food. *Ecology,* 29: 54. 1948.

148. FÄNGE, R., K. SCHMIDT-NIELSEN AND H. OSAKI. The salt gland of the herring gull. *Biological Bulletin,* 115: 162. 1958.

149. FÄNGE, R., K. SCHMIDT-NIELSEN AND M. ROBINSON. Control of secretion from the avian salt gland. *American Journal of Physiology,* 195: 321. 1958.

150. FRINGS, H., A. ANTHONY AND M. W. SCHEIN. Salt excretion by nasal gland of Laysan and Black-footed albatrosses. *Science,* 128: 1572. 1958.

151. FRINGS, H., AND M. FRINGS. Observations on the maintenance and behavior of Laysan and Black-footed albatrosses in captivity. *Condor,* 61: 305. 1959.

152. HIATT, E. P., AND R. B. HIATT. The effect of food on the glomerular filtration rate and renal blood flow in the harbor seal (Phoca vitulina L.). *Journal of Cellular and Comparative Physiology,* 19: 221. 1942.

153. IRVING, L. Respiration in diving mammals. *Physiological Reviews,* 19: 112. 1939.

154. LADD, M., RAISZ, L. G., CROWDER, C. H., JR., AND L. B. PAGE. Filtration rate and water diuresis in the seal, Phoca vitulina. *Journal of Cellular and Comparative Physiology,* 38: 157. 1951.

155. SCHMIDT-NIELSEN, B. The resourcefulness of nature in physiological adaptation to the environment. *The Physiologist,* 1: 4. 1958.

156. SCHMIDT-NIELSEN, K. AND R. FÄNGE. The function of the salt gland in the brown pelican. *The Auk,* 75: 282. 1958.

157. SCHMIDT-NIELSEN, K. AND R. FÄNGE. Salt glands in marine reptiles. *Nature*, 182: 783. 1958.
158. SCHMIDT-NIELSEN, K., C. B. JØRGENSEN AND H. OSAKI. Extrarenal salt excretion in birds. *American Journal of Physiology*, 193: 101. 1958.
159. SCHMIDT-NIELSEN, K., AND B. SCHMIDT-NIELSEN. Water metabolism of desert mammals. *Physiological Reviews*, 32: 135. 1952.
160. SCHMIDT-NIELSEN, K. AND W. J. L. SLADDEN. Nasal salt secretion in the Humboldt penguin. *Nature*, 181: 1217. 1958.
161. SMITH, H. W. The composition of urine in the seal. *Journal of Cellular and Comparative Physiology*, 7: 465. 1936.
162. VIMTRUP, BJ., AND BODIL SCHMIDT-NIELSEN. The histology of the kidney of kangaroo rats. *Anatomical Record*, 114: 515. 1952.

Observations on the osmotic concentration in the blood and urine of whales and other marine mammals are cited by Krogh {65}, Smith {161}, and Prosser *et al.* {54}. Clarke and Bishop {147} discuss the nutritional aspects of plankton, and point out that when the sea water is squeezed out by hand, the resulting moist material contains 70 per cent as much chloride as sea water; when metabolic water is added, the plankton appears to afford sufficient water to permit the excretion of metabolic products in a urine no more than four times as concentrated as the blood. The margin of available water is, however, a narrow one. The fate of the ingested magnesium and sulfate is undetermined.

That the seal-like pattern of water conservation may be remotely related to the carnivorous habit is suggested by the fact that the dog shows a qualitatively similar but quantitatively much less marked relation between protein diet and filtration rate {62, p. 470 f}, but it seems more likely that the pattern in the seal is related to the fact that it is a diving mammal. The diving reflex is unique to the diving mammals, and oxygen-lack has no such action in other mammals; in man it induces, on the contrary, a slight to moderate increase in renal blood flow and filtration rate {62, p. 441}.

Forster {115} has shown that if one closes the nostrils of the ordinary white rabbit, the animal rapidly asphyxiates and will die in convulsions without reduction in the renal circulation; but if it is presented with a noxious olfactory stimulus (whiffs of cigarette smoke), a reflex entirely similar to the diving reflex is elicited in which the renal circulation is sacrificed. Though cigarette smoke is an artificial stimulus, it is conceivable that some comparable stimulus acts on the rabbit when it is in the depths of its burrow, and possibly serves to conserve oxygen. The presence of some such reflex may explain the great lability of the renal circulation in this species {62, p. 535 f}.

XII. MAN

163. *Cold Spring Harbor Symposia on Quantitative Biology. Volume XV. Origin and Evolution of Man.* The Biological Laboratory, Cold Spring Harbor, L. I., New York, 1950.

164. Commemoration of the Centennial of the Publication of *The Origin of Species* by Charles Darwin. *Proceedings of the American Philosophical Society,* 103: 153–319. 1959.

165. Herrick, C. J. *An Introduction to Neurology.* W. B. Saunders Company, Philadelphia, 5th ed., 1931.

166. Howells, W. W. Origin of the human stock. Concluding remarks of the Chairman. {See 163, p. 79.}

167. Krogman, W. M. Classification of fossil men. Concluding remarks of the Chairman. {See 163, p. 119.}

168. Morton, D. J. Human origin. Correlation of previous studies of primate feet and posture with other

morphologic evidence. *American Journal of Physical Anthropology*, 10: 173. 1927.

169. ROBINSON, J. T. The nature of *Telanthropus capensis*. *Nature*, 171: 33. 1953.

170. SCHULTZ, A. H. Origin of the human stock. The specializations of man and his place among the catarrhine primates. {See 163, p. 37.}

171. SIMPSON, G. G. Some principles of historical biology bearing on human origins. {See 163, p. 55.}

172. SMITH, G. E. Essays on the Evolution of Man. Oxford University Press, London: Humphrey Milford, 2nd ed., 1927.

173. STRAUS, W. L. The riddle of man's ancestry. *Quarterly Review of Biology*, 24: 200. 1949.

174. ZUCKERMAN, S. Taxonomy and human evolution. *Biological Reviews*, 25: 435. 1950.

Among the many books appearing in 1959, marking the centennial of the publication of *The Origin of Species* by Charles Darwin, the group of ten essays sponsored by the American Philosophical Society {164} is particularly noteworthy. Every essay has been prepared by a writer of outstanding competence in his field of evolution. In the discussion of *Australopithecus* I have closely followed the first of these essays, 'The Crucial Evidence for Human Evolution,' by Le Gros Clark.

For other discussions of the origins of man and the significance of the australopithecines, reference may be made to Howells {166}, Krogman {167}, Schultz {170}, Simpson {171}, Gregory {33, p. 480}, Robinson {169}, Straus {173} and Zuckerman {174}.

G. Elliot Smith's {172} older volume on the evolution of man is a classic that can be recommended to the non-technical reader. The 1927 revision is to be preferred to the original print of 1924. This work, of course, antedates the discovery of *Australopithecus*.

The role of the foot in the evolution of the bipedal habitus is discussed by Morton {168}.

Léon Fredericq is quoted by Cannon {5, p. 21}.

The concept of protoplasm as basically a physical-chemical mechanism having the character of a self-integrating, self-restoring, self-centered system was developed some years ago by the writer in *Kamongo* {82}.

Herrick {165} is mentioned here because, despite its age, his book remains one of the most readable treatises on the nervous system. Reference may also be made to Romer {15} for the comparative anatomy of the vertebrate brain.

XIII. CONSCIOUSNESS

175. COBB, STANLEY. *Foundations of Neuropsychiatry.* Williams & Wilkins Company, Baltimore, 5th ed., 1952.
176. HALDANE, J. B. S Human evolution: past and future. {See 22, p. 405.}
177. HUME, D. *A Treatise of Human Nature.* Reprinted from the original edition in three volumes and edited, with an analytical index, by L. A. Selby-Bigge. Clarendon Press, Oxford, 1896 (1928).
178. KOEHLER, O. The ability of birds to "count." *Bulletin of Animal Behavior* No. 9: 41. 1951.
179. LASHLEY, K. S. Persistent problems in the evolution of mind. *Quarterly Review of Biology,* 24: 28. 1949.
180. NOBLE, R. C. *The Nature of the Beast. A Popular Account of Animal Psychology From the Point of View of a Naturalist.* Doubleday, Doran and Company. 1945.
181. PAGET, SIR J. *Memoirs and Letters of Sir James Paget.* Edited by Stephen Paget. Longmans, Green and Company, New York, 1902.
182. SEABORG, G. T. The transuranium elements. *Endeavour,* 18: 5. 1959.
183. SHERRINGTON, SIR C. *Man on His Nature.* Macmillan Company, New York, 1941.

184. SMITH, H. W. *The End of Illusion*. Harper and Brothers, New York, 1935.

185. SMITH, H. W. Organism and environment: dynamic oppositions. In *Adaptation*. Edited by John Romano. Cornell University Press, Ithaca, 1949.

186. SMITH, H. W. The biology of consciousness. In *The Historical Development of Biological Thought*. Edited by C. McC. Brooks and P. F. Cranefield, Hafner Publishing Company, New York, 1959.

187. SPERRY, R. W. Neurology and the mind-brain problem. *American Scientist*, 40: 291. 1952.

188. STONE, C. B., Editor. *Comparative Psychology*. Prentice Hall, 3rd ed., 1951.

189. TYNDALL, JOHN. *Fragments of Science*. Volume 2. D. Appleton and Company, New York, 1898.

An interesting discussion of the inheritance of musical talent and its development early in childhood is presented by Scheinfeld {23}.

The agility of Miss Shakuntala Divi in mental arithmetic is described in an article entitled 'Numbers Game,' appearing in *Time* magazine, July 14, 1952, p. 49.

The matter of intelligence in animals is discussed by Prosser *et al.* {54} and Koehler {178}. An excellent popular account of animal behavior is that of Mrs. Noble {180}. The writer is indebted to Dr. Henry W. Nissen of the Yerkes Laboratories of Primate Biology for comments on the ability of the chimpanzee to count. A comprehensive survey of comparative psychology is presented in the volume edited by Stone {188}.

The discussion of consciousness presented here is an extension, philosophically, of the view presented in earlier books and essays by the writer {82, 184, 185}, and relies heavily on the work of many neurophysiolgists, and particularly that of Lashley {179}. Some of the views expressed here on the significance and evolution of consciousness have been discussed at greater length by the writer in an essay which treats the subject historically from Descartes to T. H. Huxley {186, pp. 109–136}.

The quotations from Tyndall are taken from that author's Belfast Address {189, pp. 190, 195}, and Hume's remarks are quoted from a reprint of his Treatise {177, pp. 634–636}.

In any discussion of the time-binding quality of consciousness, it may be recalled that the cerebral cortex in the mammals is an outgrowth of the 'nose brain,' the most anterior portion of the brain stem of the lower vertebrates. G. Elliot Smith {172, p. 190} has pointed out that the sense of smell is, so to speak, 'the cement that binds into one experience all the events that intervene between anticipation and consummation. By conferring upon consciousness in the lower animals the element of cohesion it makes possible the ultimate appreciation of time and space, the ability to look backward and forward, to remember and to anticipate. In these possibilities lies the germ of the aptitude to learn from experience and to attain skill in modifying behaviour in accordance with the changing conditions of the outside world.'

Among the vertebrates up to the primates, the sense of smell is dominant: the world presents itself as a flux of fluid, interpenetrating, shifting, but always highly significant odors, whether water- or air-borne. However, with the development of binocular vision in the primates and, subsequently, increased emphasis on the tactile sense in man, the olfactory sense lost its priority. No better example can be afforded than the matter of sexual stimuli: whatever the efforts of the perfume manufacturers, the human male is governed in sexual matters primarily by the senses of sight and touch.

Haldane {176} has expressed a view related to that presented here:

'We use our brains for thinking, but it is a mistake to suppose that the brain is primarily a thinking organ.

. . . The human brain has two superanimal activities, manual skill and logical thought. Manual skill appears to be the earlier acquisition of the two, and the capacity for language and thought has grown up around it. If we bred for qualities which involved the loss of manual ability, we should be more likely to evolve back to the apes than up to the angels.'

INDEX

THE NATURAL HISTORY LIBRARY

ANCHOR AND DOLPHIN
BOOKS OF RELATED INTEREST

Printed in the USA
CPSIA information can be obtained
at www.ICGtesting.com
LVHW020232240124
769826LV00003B/36